ABOUT THIS SERIES . . .

The ultimate responsibility for determining the fate of this nation rests with its citizens. Given that responsibility, citizens must inform themselves so they can make wise decisions. For the citizenry to do so requires, however, that sources of the facts and knowledge essential to the decision-making process must share in the responsibility by making information readily accessible to the public.

Because of the complexity and seriousness of the problems facing our nation today, special public communications efforts are needed more than ever before. This series represents one such effort by a basic industry, the nation's electric utility companies, through the Edison Electric Institute.

"Decisionmakers Bookshelf" seeks to provide to the public important discussions and reasoned viewpoints on national policy problems related to energy.

ABOUT EEI . . .

Edison Electric Institute is the association of America's investor-owned electric utility companies. Organized in 1933 and incorporated in 1970, EEI provides a principal forum where electric utility people exchange information on developments in their business, and maintain liaison between the industry and the Federal Government. Its officers act as spokesmen for investor-owned electric utility companies on subjects of national interest.

Since 1933, EEI has been a strong, continuous stimulant to the art of making electricity. A basic objective is the "advancement in the public service of the art of producing, transmitting and distributing electricity and the promotion of scientific research in such field." EEI ascertains factual information, data and statistics relating to the electric industry, and makes them available to member companies, the public, and government representatives.

WORLD ENERGY

Will there be enough in 2020?

Edited and Condensed by William Nesbit

WORLD ENERGY: LOOKING AHEAD TO 2020
ISBN 0—931032—06—7

1111 19th Street, N.W.
Washington, D.C. 20036

Printed in the United States of America at
Sheridan Printing Company, Inc., Alpha, NJ

PREFACE

The Conservation Commission, comprised of members from the western world, the centrally planned economies, the developing countries, the United Nations and the World Bank, was accorded excellent cooperation from the many organizations to which it turned for the information contained in this report.

The decisions that the world makes now will determine the energy future, particularly that of the developing nations. We of the Commission hope that the results of our study will contribute to the formulation of those decisions. The efforts of the Edison Electric Institute to provide a wider distribution of the essence of the report are very greatly appreciated.

JOHN R. KIELY
Chairman, Conservation Commission

ACKNOWLEDGMENT

This publication is a condensation of the book *World Energy: Looking Ahead to 2020.*

World Energy: Looking Ahead to 2020 is a technical report by the Conservation Commission of the World Energy Conference published in 1978 by IPC Science and Technology Press of the United Kingdom and the United States. It is presented here in abstract in order to enhance its appeal to the educated public and to gain a broader distribution for the timely and important information it contains.

The Edison Electric Institute wishes to thank the Conservation Commission, the World Energy Conference, and IPC Science and Technology Press for permission to publish this information in this edited and condensed form, and for their cooperation in the venture.

Copies of the full report, *World Energy: Looking Ahead to 2020,* (ISBN 0 902852 93 0) are available from the publisher: IPC Science and Technology Press, IPC House, 32 High Street, Guildford, UK, GU1 3EW; and 205 East 42nd Street, New York, New York 10017.

CONTENTS

ABOUT THE BOOK

Since its inception in 1924, the World Energy Conference has been actively engaged in assembling and publishing estimates of the world's primary energy resources. As early as the 1960s, in light of what appeared to be the development of potential shortfalls in energy availability, the need was recognized for a long-term study relating world energy resources to demand and conservation. Consequently, in 1972, the World Energy Conference envisioned a Conservation Commission which might conduct such a study.

This Conservation Commission was formally established in 1975. Its objectives were to evaluate, for the period of 1985 to 2020, future primary energy supply prospects, and to determine the possible courses of action directed toward overcoming potential shortfalls in energy availability—with particular attention being given to the impending decline in oil and gas supplies. The commission's report discusses in detail the potential supply of energy from all primary energy sources, the impact of possible conservation measures, and likely energy demand scenarios for the period to 2020. In addition, the report is worldwide in scope including practically all of the nations of the world: developed, less-developed, and centrally-planned.

In compiling its information, the commission made arrangements for individual nations with particular expertise to make the basic resource study. The French Petroleum Institute prepared the Petroleum Resources and Production Study. The Central Coal Research & Development Institute of Essen, Germany, prepared the coal study. The International Gas Union studied the gas supply to the year 2000; and the American Gas Association prepared the overall report to the year 2020.

The Nuclear Study was prepared in Canada. The Department of Energy and Mines, which had participated with the International Atomic Energy Agency in uranium resource studies, prepared the uranium availability study, and Atomic Energy of Canada, Ltd. (AECL), prepared the nuclear generation study.

The Electric Power Research Institute of Palo Alto, Ca., prepared the estimate for renewable resources: solar; geothermal; wind; biomass; wave; tidal; ocean thermal gradient; and fusion. Ellis Armstrong, formerly head of the U S. Bureau of Reclamation, was retained to prepare the hydro power study in which he had substantial assistance from the World Bank.

A group in Finland, with participation by several commission members, prepared the detailed conservation study, while the Energy Centre at the Cavendish Laboratory of Cambridge University was retained to develop basic energy demand studies.

Each of the basic studies was then carefully reviewed by panels of experts from around the world, with the draft conclusions presented for a critique at round table forums at the World Energy Conference meeting at Istanbul in late 1977. The basic studies were revised to take into account this worldwide critique, and were finalized in early 1978. The commission, using these basic documents and taking cognizance of other energy studies, then prepared its final report which was presented to the assembled nations at Banff, Canada, in late September of 1978.

INTRODUCTION

The world in the year 2020 will be a far different place than it is today, and one of the most important determinants of that world will be the availability of adequate energy supplies at a reasonable cost.

According to the Conservation Commission of the World Energy Conference in its 1978 report, *World Energy: Looking Ahead to 2020,* this energy can be available, and at a reasonable cost, if the world reacts appropriately during what the commission sees as an unavoidable and imminent world energy "transition period."

The emergence of this energy transition period is a key finding in the report, evolving from the fact oil and natural gas production is expected to begin leveling off around 1990 to 2000 and declining thereafter. New supply must be available at that time to meet demand, but—and for the first time in the history of the world—there may not be a readily available and readily acceptable low cost energy resource standing by, a resource prepared to take over as coal did for wood in the 19th century, and as oil and gas did for coal in the 20th century. How the world responds during this transition period, the commission stresses, will have much to say about its energy future.

In the commission's view, if this transition can be successfully negotiated, world energy supply in the year 2020 can be sufficient to meet a demand of 1000 exajoules. This total represents a compounded energy growth rate of 1.3% a year for developed nations, 5.7% a year for less-developed nations, and 3.3% a year for centrally-planned nations.

Energy growth at this rate would support an annual economic growth rate of 4.2% for developed nations and 5.7% for less-developed nations.

However, and as the commission points out, energy projections and policy necessarily involve elements of ideology and of ethics, as well as of trade-offs between different political objectives—that is to say issues of a political nature which cannot be addressed in a technical report.

Moreover, in each particular country an appropriate energy policy has to be adjusted to the local circumstances, taking social, political, and economic factors into account, while considering what labor, capital, and indigenous energy resources are available for development. Again, these are factors which could not be addressed—factors of potentially decisive importance to the projections.

In its report, the commission touches a number of key points regarding the future energy picture. Some of these (which are discussed further in the chapters which follow) are as follows:

- Of the 1,000 exajoules of energy potentially available in 2020, nuclear energy (fission) is projected to supply about 31%, coal 26%, conventional natural gas 12%, conventional oil 11%, renewable resources 10%, hydraulic 6%, and unconventional gas and oil 4%.
- Even assuming an immediate breakthrough in solar, fusion, or geothermal "alternate energy" resources, these resources cannot be expected to account for more than about 10% of world primary energy production by 2020.
- The world can conservatively expect an eightfold increase in electrical generating capacity by 2020, representing an average growth rate of 4.4% a year. By 2020, in fact, it is expected that approximately half of the primary energy produced in the world will be committed to electricity generation.
- If there is prompt response in implementing energy supply and conservation programs it can be expected that developed countries—although they will be dependent on imported oil over the entire period 1985 to 2020—will become a net exporter of energy after 2007. This shift to the role of energy exporter coincides with the beginning of a projected energy deficit in the less-developed nations, a deficit which could be supplied by coal and uranium from developed nations. With 10 years delay in supply and conservation programs, however, prior to 2020 there will occur a rather sudden world oil shortage which will rapidly transform into a generally and gradually worsening energy squeeze.
- There seems to be a trend toward "energy saturation" in industrial nations, a trend in which there is an apparent decreasing requirement for additional energy with increasing gross national income. This is important because if the relationships which exist today between energy and gross national income in the developed and less-developed nations of the world were to continue unchanged, the projected increase in energy supply would not be sufficient to support the higher rates of economic growth desired by the less-developed nations.
- About 50% of the energy which would otherwise have been used by 2020 may possibly be saved through conservation. Technological development and improved energy conversion and utilization efficiencies are expected to account for approximately 30% of this total. In addition, further conservation, possibly 17%, can come about through

structural changes in the more mature economies (i.e., through saturation in demand for car transportation, reduced utilization of energy-intensive industries, and significant changes in domestic energy requirements).

- A continuous increase in energy prices could trim world demand for energy that would otherwise have been used by the year 2020 by some 400 exajoules, a total more than the world's current production of primary energy.

Despite the fact the commission report is a technical one, it does propose actions which it feels policy-makers must take into account. These include:

- The problem is an urgent one and it is necessary for the world to come to grips with it without further delay.
- There is no easy way out of the problem, and even efficiency in energy utilization—which the commission feels must be pursued as a top-priority item—may not be able to improve on (or even to maintain) past trends.
- The production of primary energy must be promoted with comprehensive programs designed to develop indigenous resources.
- Cooperation, internationally between governments, and between governments and industry, is required regarding the transfer of technology and capital resources, in developing international trade in energy commodities and in bringing about financial and fiscal reforms geared to promoting indigenous energy forms.
- There must be a more equitable distribution of primary energy, and political leaders should direct development towards the productive use of energy in less-developed countries and towards a saturation of energy demand in the highly industrialized nations.
- Conservation must be diligently applied, possibly using legislative and fiscal means.

GLOSSARY

Income and Price Elasticities: The rates at which energy demand grows with increases in national income and energy prices.

Exajoule(s): A unit of energy measure representing approximately 160 million barrels of oil.

Energy Saturation: A trend among industrialized nations which indicates that there is a decreasing requirement for additional energy with increasing gross national income.

Conventional Petroleum: That crude oil together with natural gas liquids which is economically recoverable using current technology under current attractive economic conditions. In practice, it is the petroleum now produced both onshore and offshore at water depths of less than 650 feet.

Unconventional Petroleum: Oil deposits which will require exploration and exploitation technology not yet fully developed, so that oil production from them appears marginal today.

Energy Resources: The global aggregate of energy-bearing, nonrenewable minerals and renewable energy forms which can with some degree of confidence be assumed to be exploitable eventually.

Energy Reserves: That fraction of resources which can be recovered by means of reasonable extensions of today's recovery technologies at costs that are expected to be economic.

Gigatonne: A unit of energy measure equivalent to 1 billion tonnes and approximately 7 billion barrels of oil.

Tonne: The equivalent of 1,000 kilograms, or 2,204.6 pounds.

Terawatt: A unit of energy measure. Energy supplied at the rate of one terawatt over a period of one year is equal to 31.54 exajoules.

Biomass: The mass or amount of living organisms in a particular area. Used here to refer to firewood, urban and agricultural wastes, animal dung, plant life, and the like, which can be used for their energy content.

1. DEMAND

In its report, the Conservation Commission presents three separate projections for energy demand to the year 2020: the first, an extrapolation of historical trends in energy use; the second, a specially-commissioned study conducted for the commission by the Energy Research Group, Cavendish Laboratory, Cambridge University, Great Britain, and called "The Cavendish Report"; and the third, an "Alternative Demand Scenario," in which the commission projects how energy demand and economic growth are affected by income and price elasticities not considered in the Cavendish Report.

The approach endorsed by the Conservation Commission is the alternative scenario which is in essence an extension of the Cavendish group's work.

Taken in order, the most straightforward of the projections is the extrapolation of past demand. While the commission feels the range in demand which results from extrapolation—a high of 2100 exajoules and a low of 700 exajoules (*Table 1*)—is much too broad, it does point out that the technique is helpful in delineating a range outside of which world energy demand is not likely to fall.

The Cavendish Report was commissioned in March of 1976 and is described as a work of pioneering nature. Briefly summarized, the report uses an energy demand model which relates energy demand to economic growth and energy price (it is generally accepted that the use of energy is related to economic activity) through econometric equations which employ income and price elasticities. In this model, important assumptions are made about the anticipated growth of the world economy, and anticipated increases in world energy prices through the years. Assumptions of world economic growth were made at a "low" level of 3% and a "high" level of 4.1% averaged over the period 1975 to 2020, and energy demand projections were then related to these figures. (*Table 2.*)

Table 1 World primary-energy demand projection based on different historical periods

Historical period	*Annual rate of growth over period, % per annum*	*Primary-energy demand by extrapolation to the year 2020, exajoules*
1860 - 1975	2.0	700
1925 - 1975	3.3	1306
1933 - 1975	4.1	1918
1960 - 1975	4.3	2100

When the Cavendish team explored the possibility of an even higher rate of economic growth in less-developed nations—a rate higher than that factored into its 4.1% total world economic growth assumption—it concluded that even when high conservation was assumed in the less-developed nations, and low growth was assumed in developed nations, estimated potential supply was exceeded before 2020.

The commission's alternative scenario provides broader insights into energy-demand/economic-growth relationships by considering and incorporating changes over time in income and price elasticities. (*Table 3.*) In the commission's view, there is no reason for the elasticities to remain the same over the next 50 years. On the contrary, they can be expected to vary with time and conceivably the changes could become large.

Specifically, the commission felt the values chosen for income elasticity should be lower for developed countries and higher for less-developed countries, and that as these economies grow the elasticities would tend to decline. In addition, in this scenario significantly higher world economic growth rates are allowed, particularly for the less-developed part of the world, but also for the developed region: a rate of 4.2% a year for developed nations, and 5.7% a year for the less-developed.

With these and other revised assumptions, together with an annual increase in energy demand of 3.3% for the centrally-planned economies, energy demand for the year 2020 is seen to be 1000 exajoules—a level world energy supply should, in the commission's view, be able to meet at that time.

Table 2 Alternative assumptions for gross world product (GWP) growth rates in Cavendish report, %

	OECD nations	*Centrally planned economy nations*	*Developing nations*	*World*
High growth	3.7	4.5	5.3	4.1
Low growth	2.8	3.2	3.8	3.0

Table 3 Alternative Scenario demand projections

	Primary energy demand, exajoules			
	OECD	*Centrally planned economies*	*Developing nations*	*World*
1972	150	66	27	243
1980	178	86	46	310
1990	212	120	86	418
2000	242	167	152	561
2010	262	233	253	748
2020	278	325	397	1000

Note: Only commercial energy is shown in the projections.
1EJ = 10^{18}J

Variables in Assessing Demand

Some important points concerning future world energy demand pinpointed by the commission and not to be overlooked include the following.

• The Cavendish energy demand model has no self-regulatory mechanism, that is to say there is no feedback from the energy sector into the rest of the economy. This is acceptable as long as the energy share (i.e., the role of energy) in the economy is small. However, if—as the Cavendish Report suggests—energy availability becomes a major constraint on economic growth, the situation would change significantly. Very little is known, in fact, about the nature of the feedback mechanisms and the possibilities of substitution between energy and non-energy (e.g., capital and labor) inputs into the economy.

• In the Cavendish Report the price of energy is related to the price of oil according to a certain mathematical formula, and no forecast of future real energy prices is made. While the values chosen for the price of oil are based on estimates approved by the Conservation Commission, the commission still points out that the coupling of oil and other energy prices is open to question. This is especially significant in that one major conclusion of the commission's study is that the supply of conventional oil will begin to decline around 1990 and gradually be phased out and reserved for premium uses (transportation and chemical feedstocks, as examples). Under this circumstance, it seems reasonable, as the commission notes, to assume that the price of energy will eventually "decouple" from the price of oil, and that the price of oil would then have but a small impact on overall energy prices.

• Another significant variable broached by the commission and affecting its projections is the question of whether societies are going to consume ever-increasing amounts of energy as they become more and more wealthy. In its study, however, the commission noted a trend toward "energy saturation," a pattern in which there seemed to be a decreasing requirement for additional energy with increasing gross national income among industrialized nations. By the year 2020, in fact, it appears that the ratio of the unit energy increase to the unit income increase might decline to one-half of what it is today. In the case of less-developed nations now in the take-off stage of economic growth, considerably larger energy input is required in support of economic growth, and—during the initial phase of rapid industrialization—the potential for energy conservation is probably smaller than in the more mature, energy-intensive industrialized societies. Commission demand projections for the less-developed nations are such that by the year 2020 these nations will require slightly greater ratios of energy per unit of gross national income than applies to developed nations today.

The Cavendish H5 (high-growth energy constrained) scenario for demand

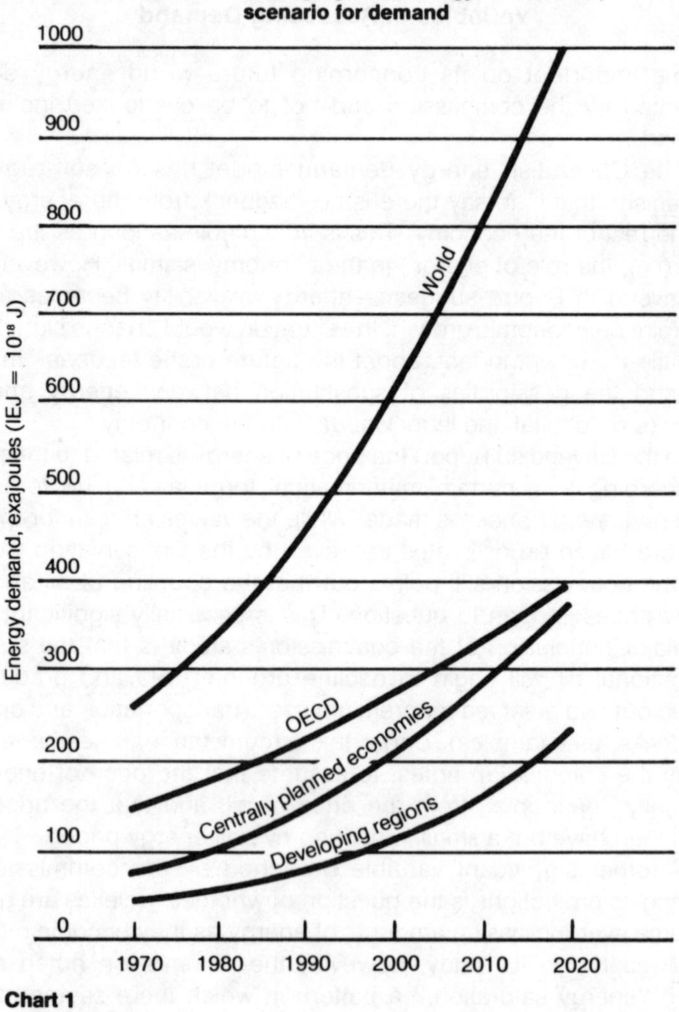

Chart 1

Relationship between energy consumption and economic activity.

(Based on an original in Scientific American, Vol. 224, No. 3 (1971)

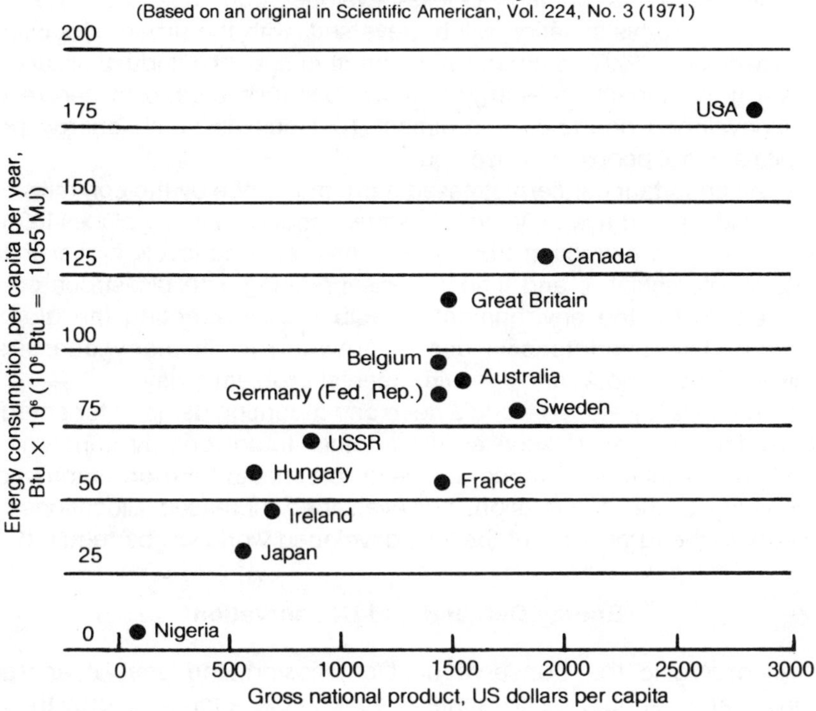

Chart 2

In addition, if the relationships between energy and gross national income which exist today in the developed and less-developed parts of the world were to continue unchanged, the projected increase in energy supply would not be sufficient to support the higher rates of economic growth which are desired by the less-developed countries. In some way, therefore, the ratio of energy to total national income in more advanced countries must decline if the energy supply projected as attainable is to be sufficient to support satisfactory improvement in material conditions, especially for those who have less. The commission feels this decline will come about naturally, as economic development proceeds toward energy saturation. If required, however, it may have to be induced by active political measures.

• An additional, important variable in projecting world energy demand is world demography. The world's present, dramatic population increase is not expected to level off significantly until well beyond 2020, and by the year 2000 the less-developed part of the world is expected to represent nearly 4 billion people—or more than 60% of the world's population at that time. By 2020, this share is expected to rise to nearly 70% of a global population of 8 billion.

On top of this, a continuation of the current massive urbanization process is expected. While today some 60% of the world's population lives in rural areas (primarily in less-developed countries), by 2020 it is expected that this situation will be reversed, with the urban population approaching 60%. This urbanization, another aspect of industrialization, has a great impact on energy demand because urbanized people in less-developed countries use almost 10 times as much energy per capita as rural people in those countries.

- An important concern stressed more than once by the commission is the widespread use of firewood as fuel. Indeed, in many places in the world firewood extraction from forests has reached levels higher than natural reforestation, and timber is disappearing with disastrous consequences for the environment, possibly even affecting the global climate. This overintensive use of firewood is, in many quarters, considered the most serious environmental problem today.
- The less-developed world's need and mounting demand for energy for agricultural use is seen as yet another future energy concern. If modern agricultural methods are used, prospects here are improved, according to the commission, but even then increased allocations of energy to the rural areas of the less-developed world will be required.

Energy Demand and Conservation

According to the Conservation Commission, the goal of energy conservation should be to achieve acceptable economic growth (in terms of gross national product, or GNP) with a minimum increase in total energy consumption—that is to say, to use energy resources more efficiently, causing minimum change to existing, desired lifestyles.

Reaching this goal does not necessarily mean that the ratio between consumption and gross national product has to decrease in all sectors under all circumstances, the commission points out. In countries where electrification is in an early stage of development, as an example, electricity demand obviously will grow faster than gross national product to fuel an appropriate growth in the economy.

Possibly the strongest force affecting energy consumption is energy pricing, which not only encourages conservation but also the development of resources. Other factors include "doing better" (i.e., technological improvements), "doing with something else" (i.e., substituting new resources of energy for diminshing resources), and "doing with less" (i.e., changing lifestyles). These options are not mutually exclusive and may be pursued simultaneously.

As regards "doing better," the commission reports that about 50% of the energy that would otherwise have been used by 2020 may possibly be saved. Technological development and improved energy conversion and utilization efficiencies are expected to account for approximately 30% of this total, and this type of saving can be achieved in all sectors of the world's economy, it is said. In addition, further conservation,

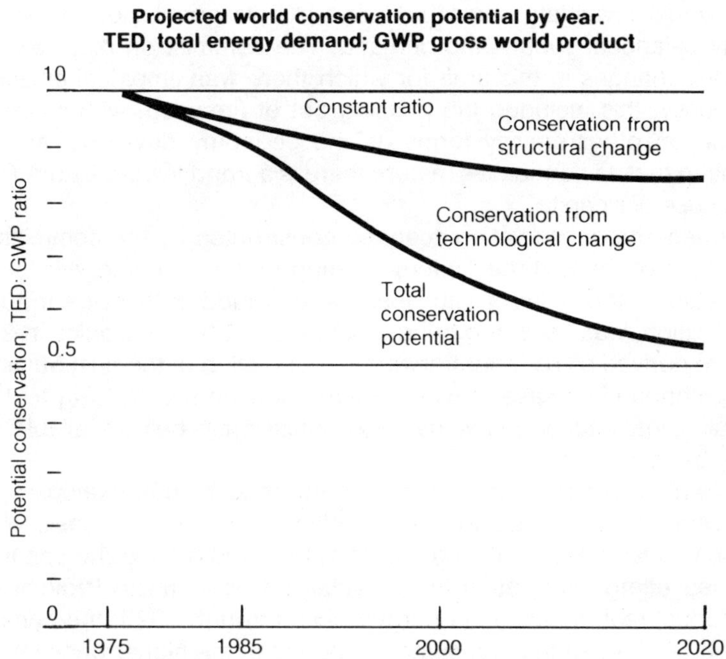

Chart 3

possibly 17%, can come about through structural changes in the more mature economies, changes such as saturation in the demand for car transportation, reduced utilization of energy-intensive industries, and significant changes in domestic energy requirements. (*Chart 3.*)

As regards "doing with something else," the commission points to one of the most important findings of its study: the fact that the production of conventional oil and natural gas is expected to begin leveling off around 1990 to 2000, and declining thereafter. When this occurs, the commission argues, conventional oil and gas will have to be reserved for uses for which they are uniquely suited (i.e., transportation and chemical feedstocks), and new resources developed to take their places. However, while candidate new resources have been identified, it is expected that 50 to 60 years will be required for them to penetrate the energy market to any significant degree. As a result of this long phase-in period, substitution efforts must be undertaken immediately.

The question of "doing with less," the commission says, is essentially a political one which in a technical report cannot be fully addressed. Involved are such market factors as decisions on price policy and taxes, regulatory measures, standards and quotas, as well as such non-market factors as subsidies, grants, and loans as incentives for energy-saving investments. Deliberate or politically-enforced changes in lifestyle aimed at lowering the average level of energy consumption, the commission says, would invalidate some but not all of its conclusions regarding energy balances, supply and demand. The commission did, however, consider changes in this area for which there was empirical evidence. Specifically, this included the phasing out of firewood, with increased reliance on other energy forms as the economy develops, and the indication that as economies mature there is a trend toward saturation in energy use per capita.

In an energy conservation scenario constructed by the commission, the impact of these three factors—"doing better," "doing with something else," and "doing with less"—combined with price-induced conservation, was shown to bring a demand of 2000 exajoules in 2020 (a figure derived by a straightforward extrapolation of the past 50 years' average annual increase in world energy consumption) down to 1000 exajoules, that total projected by the commission as being available on a supply basis in 2020.

This was accomplished by reducing the total by 600 exajoules with improvements in energy utilization efficiencies (on the basis of an historical, average annual increase of 1.4%), and adding the impact of predicted energy saturation in industrialized economies. Projecting a continuous, real increase in energy prices through 2020, the commission then estimated that consumer response to the higher prices would trim an additional 400 exajoules off the total, bringing annual demand to the 1000 exajoule supply potential derived for 2020. This level of price-induced conservation amounts to almost twice the world's current production of primary energy, and will have a real impact on the output of goods and services.

Points on Conservation

Additional, important energy conservation points made by the commission include the following:

- While conservation is a necessary ingredient in any sensible energy policy for the future, it is not sufficient without additional measures.
- It is no longer possible to continue patterns of energy use or to use energy-consuming processes which were economic in the recent era of abundant and inexpensive energy supply.
- In light of demand projections and in view of the transitional period at the end of this century when oil production is expected to decrease, all possible efforts should be made to secure economic growth at a lower energy input per unit of production than at present.
- In most industrialized countries the ratio between the percentage growth of energy use and the percentage growth of gross national product is less than one. Since 1974, however, while economic growth in the market economies has continued at a somewhat reduced rate, energy consumption in some cases has actually decreased. This can be

attributed in part to the fact oil prices increased so sharply, as well as to newly introduced energy conservation programs.

- Even when it is assumed that less-developed countries will employ modern, energy-efficient technologies, the conservation potential for these countries is still much smaller than that for the developed countries. They are also likely to place priorities differently and, given the choice, restrict capital investment rather than energy inputs. Additionally, some of the structural change expected to take place in the economies of the developed countries implies the transfer of energy-intensive industries to less-developed countries with indigenous energy resources. While this may ease some of the pressure in developed countries, it will add correspondingly to the energy demand of less-developed nations.
- There are a number of factors limiting conservation potential. They include: fundamental limits set by thermodynamics; practical limits set by the rate of conversion; simple ignorance of already existing possibilities for saving energy; economic constraints (e.g., capital shortages, or low profitability and long payback periods for conservation measures); lack of manpower responsible for the control of energy efficiency in production processes; the fact that many industrial plants and much equipment are not sensitive to energy efficiency, that is to say their overall running costs are not greatly influenced by changes in energy costs; the fact major technical developments for conserving energy generally involve long lead times so that large effects are not likely to be achieved early; institutional obstacles between industry and utilities in some countries; and, political trade-offs (e.g., environmental) in policy making.
- Energy prices should be brought into new relations with respect to factors such as calorific value, scarcity of resources, and future energy prospects, relations reflecting world prices and aiming at more rational energy use and waste prevention.
- The dependence of economic growth on increases in energy consumption can be reduced by governments in a number of ways. Among them by: changing consumer behavior by economic and other pressures; eliminating energy waste; promoting technical developments for saving energy; substituting other factors of production (such as labor, capital, and materials for energy); substituting for existing forms of energy with more appropriate ones; discouraging production of energy-intensive and energy-using goods; intensifying publicity using mass media; introducing new legislation; promoting investigations of energy balances to define losses; introducing national standards for energy-using appliances; adopting laws for labeling equipment with information on energy consumption; subsidizing research on unconventional energy sources; subsidizing research and development directly applied to energy conservation; and promoting the recycling of energy-intensive goods.

• During this century energy has increasingly been substituted for labor as the relative prices of energy declined. With today's higher prices, however, and even in industrialized countries, some shift back to the use of labor—with the secondary advantage of providing further opportunities for employment—should be investigated.

• The three largest consumers of energy in the transportation sector are automobiles, airplanes, and trucks, each of which is also most energy-intensive using large amounts of energy per unit of work provided. Conservation measures in this area should be aimed at increasing fuel efficiency and reducing the use of fuel.

• The potential for cogeneration (the combined production of heat and power) for saving energy is believed to be very significant in the industrial sector which uses large amounts of steam, and, on the average, accounts for approximately 30% of total primary energy requirements in developed countries. Government should concentrate on establishing effective planning coordination in this area and on sweeping away administrative barriers.

• The key to successful application of conservation measures is to be found in making constructive decisions and initiating timely implementation programs.

II. SUPPLY

Studies performed for the Conservation Commission indicate that by the year 2020 the total annual primary energy supply could reach 1,000 exajoules—a total somewhat more than three times today's world supply.

Attainment of this level of supply implies a sustained 3% a year average growth in energy production over the whole of the period to 2020. Energy forms contributing to this 1,000 exajoule total will be conventional oil, conventional gas, unconventional oil and gas, coal, renewable forms of energy, nuclear energy, and hydroelectricity. (*Table 4.*)

Contributors to Supply in 2020

Although there remain substantial reserves of conventional oil, it is predicted that oil production will reach a maximum about 1990 and that thereafter its share of total primary energy will decline from approximately 40% to about 11% by 2020. This is true even though in absolute terms petroleum production is expected to decrease only slightly from

Table 4 Potential world primary energy production, exajoules (1 EJ = 10^{18} J)

Resource	*1972*	*1985*	*2000*	*2020*
Coal	66	115	170	259
Oil	115	216	195	106
Gas	46	77	143	125
Nuclear	2	23	88	314
Hydraulic	14	24	34	56
Unconventional oil and gas	0	0	4	40
Renewable, solar, geothermal, biomass	26	33	56	100
Total	269	488	690	1000

115 exajoules in 1972 to 106 exajoules in 2020. Increasingly, oil will be used in applications for which it is especially well suited, such as transportation and chemical industry feedstocks. Its other uses will increasingly be met by other energy forms.

There are also large reserves of conventional gas, reserves which are projected to support increasing production to the end of this century. At that time, estimates show this resource providing about 20% of the world's total primary energy. By 2000, however, gas production will probably begin declining, and by 2020 its share of primary energy production will be in the range of 12%.

In addition to conventional forms of oil and gas, it has been assumed that by 2020 unconventional oil and gas—from tar sands, oil shales, geopressured zones, and other resources—will provide the energy equivalent of about 4% of total world primary energy supply.

According to the commission, coal, renewable energy resources, and nuclear energy will be required to provide approximately two-thirds of the world's energy supply by the year 2020.

To fulfill its role, coal production will have to increase by more than four times over 1972, a requirement which can be met by more than adequate coal reserves if major efforts are made in the development of mines and supply infrastructure. All told, coal should represent about 26% of the world's primary energy supply by 2020.

Renewable forms of energy—solar, geothermal, and biomass—will contribute about 10% of world primary energy supply by 2020, the commission projects, providing the primary equivalent of all the electrical energy used today—an oil-equivalent equal to two-thirds present world annual oil production. Of this 10%, solar and biomass will

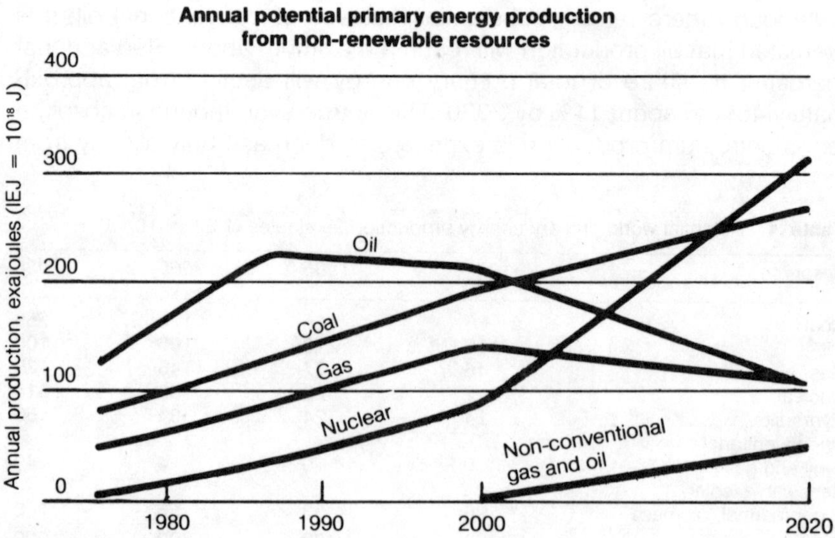

Chart 4

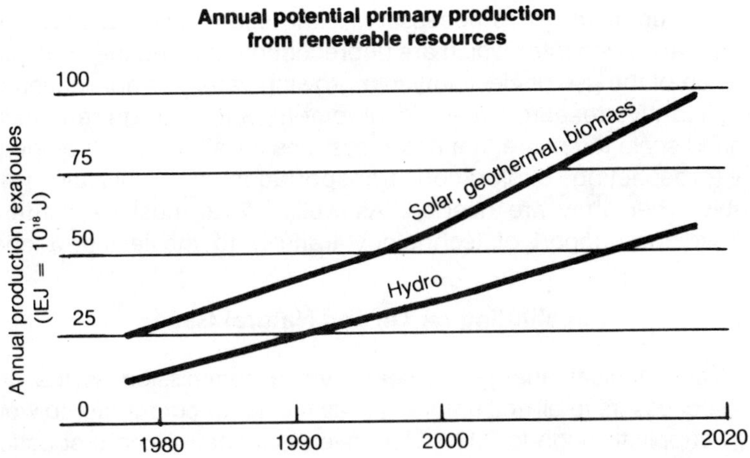

Chart 5

represent about 7% to 8%, with geothermal accounting for the remainder. When it is recognized that, in addition, the non-commercial use of wood fuel from forests—now about 7% of the world's energy supply—must decrease markedly in the same period, the implications are striking.

The largest slice of primary energy supply in 2020, some 31%, will be provided by nuclear energy from fission—which today contributes less than 1% of world primary energy production. This estimate is based on a dramatic but essential penetration of electric power supply by nuclear energy, and unprecedented development of uranium mines on a world-wide scale.

The balance of the primary energy supply for 2020—approximately 6%—will be accounted for by hydroelectricity, which should have expanded by that time to account for almost all available hydro sites.

Interestingly, even with a four decade interval until 2020, non-renewable resources are still seen as carrying the burden of energy supply: some 85% of the total in 2020. Indeed, renewable energy resources (solar, geothermal, biomass, and hydraulic) which today account for about 15% of total world primary energy production (40 of 269 exajoules), are projected to hold almost exactly the same total, 16%, in 2020 (156 of 1,000 exajoules). The burden of supply placed on the non-renewable resources, the commission reports, calls for continued improvements in technology and significantly large investments in capital and manpower.

According to the Conservation Commission, the attainment of the total level of supply required by 2020 will entail major efforts in all energy resource fields. It implies, among other things, major oil and gas exploration and development in increasingly difficult physical situations . . .the more than quadrupling of coal production. . .a doubling of nuclear power plant capacity every six years, which makes development of many new uranium mines necessary. . .the use of solar energy on a scale and at growth rates which are unprecedented. . .and the continued expansion of the use of electricity in a growing variety of applications.

Significantly, research and development will be required on an expanded scale to ensure that better systems for all parts of the energy chain (prospecting, exploration, transportation, and utilization) are available when they are needed. As well, efforts must be made to undertake the support of technology transfer to the less-developed nations.

Substituting for Oil and Natural Gas

Coal and nuclear energy are seen by the commission as the only viable successors to oil and natural gas as the major contributor to world energy supply through to 2020. Unconventional petroleum production, despite its impressive expansion, cannot fill the gap, according to the commission, and even in the most optimistic of scenarios the new energy alternatives—such as geothermal, solar, or fusion—would grow to represent only about 10% of supply by 2020.

Another consideration touched by the commission in projecting this substitution of alternate fuels for oil and natural gas is the fundamental difference between substitution patterns experienced in the past, and that substitution which must come. Petroleum and natural gas, it is pointed out, were not introduced by necessity. On the contrary, they gradually captured their market shares in a natural fashion by offering customers superior fuel at competitive prices. The coal industry adjusted accordingly, but increasing amounts of coal continued to be produced from a very respectable resource base.

The future is quite different. The production of conventional oil and natural gas is bound to decline by geological necessity. Substitution today, then, is being forced and its development with time is largely predetermined by future potential production rates.

Moreover, coal and nuclear power, the only viable supply options available during the energy transition period from 1985 to 2020, do not represent suitable substitutes for petroleum in some of its most important applications—such as transportation and petrochemical feedstock. In 1972 these "oil specific" applications accounted for nearly half of the world's oil consumption.

In its work the Cavendish team commissioned by the Conservation Commission considered this matter in some detail, concluding that total world oil demand for transport between 1972 and 2020 would increase at a much lower rate than recent trends, with "saturation," technical improvements, and vigorous measures for energy conservation contributing.

In addition, and as the commission points out, cars driven by ordinary internal-combustion engines are already running on synthetic fuels in South Africa, and on natural gas in various other locales. Further, methanol has been used in the past as an additive to gasoline and will certainly be used again. And, with only minor modifications, conventional cars can be made to operate on pure methanol.

To avoid constraints in ensuring energy for transportation, very significant developments must occur within the next decade in the area of synthetic fuel production, developments which would not be unprecedented in history. Germany, as an example, fought World War II on synthetic fuels made from coal. Between 1939 and 1945 Germany built twelve liquefaction plants and tripled its production capacity to provide, by the end of the war, half of its hydrocarbons and 98% of its aviation fuel.

The commission's conclusions on this subject are that for most of the period through to 2020 the petroleum industry will be able to satisfy the demand for energy from an expanding transportation sector. Meanwhile—since there will certainly be a strong incentive to provide some substitute for gasoline when conventional oil no longer can meet demand—it envisions that alternative fuels will be developed from coal, natural gas, tar sands, oil shale, or biomass. And, it is added, if the assumption is made that during this transition period primary-energy production will determine tolerable demand, it seems reasonable to assume that demand will adjust to whatever fuels and forms of energy are made available on the market.

Electrical Energy

A bright spot in the world energy future is in the area of electrical energy. In its estimates, the commission anticipates an eightfold increase in world electrical generating capacity, corresponding to an average growth rate through to 2020 of 4.4% a year.

By 2020, it is expected that about 5,000 gigawatts (electric) of power will be available from nuclear capacity, 2,750 gigawatts (electric) from coal-fired power plants, 1,750 gigawatts (electric) from hydroelectric power stations, and 1,500 gigawatts (electric) from oil, gas, geothermal, and solar sources. This represents a total installed capacity of about 11,000 gigawatts (electric) by 2020.

This impressive growth of electric energy is a fairly conservative projection, according to the commission. In the United States, for instance, electricity grew by 6.85% a year during the period 1960 to 1973, and electrical energy was recently projected to maintain an annual growth rate of 4.45% to 6.95% through the rest of this century. Proceeding from a much smaller base, electric power generation in India has been growing by an average of 11.5% a year since the 1950s, and is expected to continue to advance by about 10% a year during the next three decades.

By 2020, in fact, the commission expects that nearly half of the primary energy then produced will be committed to the generation of electricity. This penetration of the energy market is partly due to the rise of nuclear power as a major energy source. But also, in all regions of the world, electricity is now regarded as a high-preference energy carrier, environmentally benign and of very general applicability.

Interim Energy Supply-Demand Scenario

In its work the Conservation Commission not only tried to determine how energy supply-demand would balance *in* 1985 and 2020, but also how it would balance *between* those years. Using rates which represent upper limits for the production of primary energy, the commission constructed a supply-demand projection for this period based on the Cavendish team's high and low scenarios, and its own Alternative Scenario.

The Alternative Scenario was chosen for discussion, and it shows a successful supply of energy between 1985 and 2020 (*Chart 6*) resulting from prompt response in implementing required energy supply and conservation programs. In addition, and significantly, it shows that although developed countries will be dependent on imported oil over the entire period 1985 to 2020, they become a net exporter of energy after 2007. This shift coincides with the beginning of an energy deficit in the less-developed nations, a deficit which could be supplied by coal (and uranium) and allows the scenario to remain constant to 2020 and beyond.

With 10 years of delay in production and conservation, as opposed to "prompt response," an entirely different development pattern unfolds between the years. The developed nations now fail to meet their demand from internally-generated primary energy resources and continue to rely heavily on imported oil, and by 1990 the less-developed regions are not only unable to satisfy demand from their own rapidly-expanding economies, but also to cover the developed nations' deficit by exports. The result: no less than the onset of a rather sudden oil shortage which rapidly transforms into a generally and gradually worsening energy squeeze.

Since maximum production rates have been used in this scenario it can be shown that there is no way in which time lost in delayed response can be made up by enhanced production.

The commission's conclusions from this exercise are straightforward:

One. Unless it can be proven that the rates of expansion of primary energy used in the study are too low, the world is now more short of time than of energy in solving the problem.

Two. With sensible energy policies it should be possible to meet projected energy demand at high economic growth levels in all major regions of the world.

Three. This opportunity may be lost if decisions are delayed by only a few years. In such an event, both developed and less-developed regions will be hopelessly locked in an energy-constrained future. This conclusion, according to the commission, remains valid regardless of what values are chosen (within reasonable limits) for future energy demand in the two regions.

The Conservation Commission's Alternative Scenario

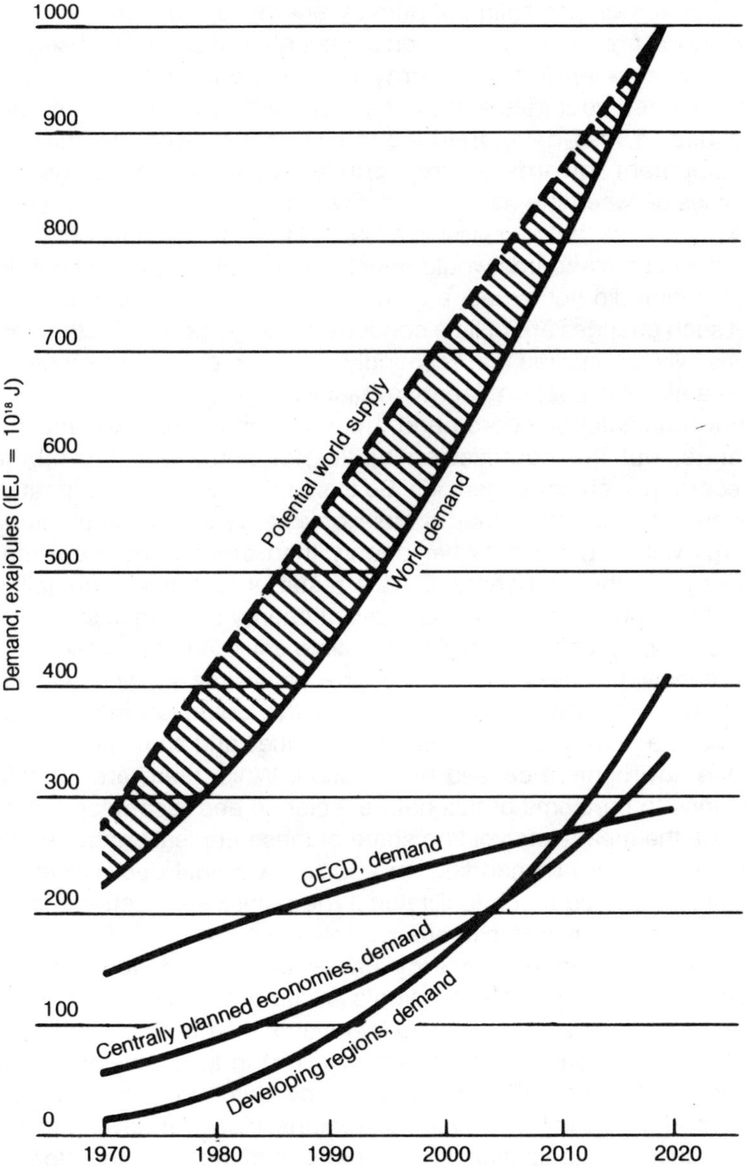

Chart 6

Some Final Points

Some additional, key points made and stressed by the commission in its study of supply strategies are as follows:

- International trade and cooperation between nations will be of paramount importance in the years to come. This cooperation must include intensified prospecting, exploration, and development in the less-developed countries, together with the transfer of technology required for the promotion of primary-energy production.

- Fiscal and legal means open to government should be used to bring about a deceleration of energy consumption growth in the more industrialized countries and to channel the flow of energy into productive use in less-developed countries, thus directing the general development towards a more equitable distribution of goods and services between the nations of the world.

- In its work the commission assumed that current environmental and anti-pollution standards would remain unchanged. This was not done to imply that it did not expect any major changes to occur in this area, or that such changes are of little concern in energy policy. On the contrary, it was fully recognized that seemingly unrelated pollution restrictions can have a major impact on energy availability.

Environmental standards and anti-pollution policies will most surely change, but the commission could not conceive of a method of forecasting such contingencies. Be this as it may, there is a great deal of evidence to the effect that in case of emergency an ample supply of energy will be given a higher priority than at least the more marginal concern for the environment. As a result, it was felt in compiling the report that proper measures to prevent energy shortages should form a part of a comprehensive and responsible environmental policy.

The individual risks and displeasures as well as social strains caused by a major energy shortage must, in the commission's opinion, be balanced against losses of pastoral amenities and some adverse effects on the environment caused by pollution. While there are no scientific solutions to problems of this nature, science and technology can often provide the means of avoiding some of these unpleasant trade-offs and make others less demanding. In addition, political decision making in this area could be much facilitated if the public were better informed of the real options available in energy policy.

- The importance of energy prices on holding down demand is clearly shown in the commission's assumption that an annual price increase of 3% a year (corrected for inflation) in developed regions, and of 2% a year (again corrected for inflation) in less-developed regions, would by the year 2020 reduce energy demand 44% in developed regions and 25% in less-developed regions. However, given the current, fairly closed organization of the petroleum market, the dominating position of the OPEC cartel, and the future prospects for petroleum production, it is highly unlikely that this steady and continuous price increase will come about through the free interplay of market forces.

• Enhanced oil production with an "accelerated" depletion rate will not result unless the price of petroleum is increased accordingly. The massive oil revenues accruing are likely to be invested within rapidly growing national economies (not necessarily in the developed regions) where the future is less intimately tied up with petroleum. That is, where alternate energy supplies already exist or are being made available. Countries that remain heavily committed to imported fuel will therefore be doubly hurt by increasing petroleum prices: first, by the terms of trade which will deteriorate, and secondly, by being unable to offer an attractive economic climate for long-term foreign investment.

• The 40 exajoules of petroleum projected to be produced by 2020 from enhanced recovery, deep sea and Arctic regions, from tar sands and oil shale, and from coal and biomass, is not likely to become available unless some financial organization is willing to supply the capital required for these financially quite risky ventures.

• Governments may consider a direct involvement in primary-energy production based on domestic resources, for instance through government-owned, corporation-operated pilot projects and demonstration plants for the production of indigenous petroleum from oil shale and tar sands, or synthetic gaseous and liquid fuels from coal. A firm political commitment to reduce the dependence on imported petroleum and natural gas with a definite program for the promotion of domestic energy supplies would help to stabilize world market prices and might therefore turn out to be a very profitable investment.

• During the period between now and 2020, disaster is the only alternative to international trade and cooperation. In this process the oil exporting countries will occupy a very prominent position, in particular the OAPEC (Organization of Arab Petroleum Exporting Countries) countries in the Middle East and North Africa as potential suppliers of some 40% of the world's petroleum production for the next half-century. Long-term working relations will have to be established and the associated flow of capital must be managed within a viable international monetary system conducive to international trade and economic growth in all parts of the world.

III. ENERGY RESOURCES

The world energy supply-demand projections and conclusions offered by the Conservation Commission are based on in-depth assessments of the world's conventional and unconventional energy resources. In this work, the commission looked at energy forms which either will be contributing to world supply by 2020, or hold so much potential for the future—perhaps even by 2020—as to merit attention today.

Included are discussions of conventional and unconventional oil, reaching to oil shales, tar sands, enhanced recovery, polar deposits, and the like; natural gas, including methane from coal beds, geopressured gas, and petroleum and coal conversion; coal resources, touching on degasification and liquefaction; nuclear energy, both fission and fusion; geothermal energy; hydraulic energy; solar resources including wave, tidal, ocean thermal-energy conversion, and biomass; and hydrogen.

Oil

While oil and tar from geological seeps have been used for heat, light, and ship construction for thousands of years, it was just over a century ago that oil was first produced on a commercial basis (in Eastern Europe and North America).

The real stimulant to oil production came in 1870 with installation of the first steam drill in the Caucasus on the western edge of the Caspian Sea. This development greatly enhanced drilling technology and marked the beginning of a burgeoning industry which grew at an astounding rate.

In its first 30 years oil production grew at a rate which doubled every five years. It doubled every 10 years between 1880 and 1930, and then again between 1930 and 1946 with development of the great oil fields in the Middle East—first Iran, then Iraq, Saudi Arabia, Kuwait, and the

United Arab Emirates. Since 1946, oil production has recorded an average growth rate of 7.3% a year, reaching a rate of 2.8 gigatonnes in 1975.

In addition to the Middle Eastern oil fields and those in Europe and North America, there are contributing fields in North Africa, Asia, West Africa, South America, Central America, Oceania, the Alaskan and Soviet Arctic areas, and the North Sea. New major discoveries are also reported annually in the USSR and the People's Republic of China.

Conventional Petroleum Resources

In estimating ultimate worldwide resources in crude oil, and the limits to future annual production, the Conservation Commission used a Delphi-type poll in which authorities from oil companies, government agencies, and consulting firms were polled on a number of questions and a consensus of these responses was formulated. A number of respondents were visited following the compilation of the poll responses to discuss the conclusions reached from the poll and to ascertain potential development in production technology.

This Delphi-type technique is used for long-term predictions of future trends in industrial, economic, social, and political conditions where it is felt that the complex elements involved in the predictions are better handled by judgment and experience than by extrapolation of current trends and other analogy techniques. In this poll the commission related recoverable petroleum resources and production capabilities to a technical cost of $20 per barrel (at 1976 prices) in 2000, and assumed that by 2000 present oil-recovery efficiency would increase from 25% to 40% in many fields.

The result showed estimated, remaining, recoverable conventional oil reserves ranging from 175 gigatonnes to 350 gigatonnes (a mean of approximately 260 gigatonnes), with the greatest weight of opinion suggesting reserves of 240 gigatonnes (*Chart 7*). In this estimate,

Remaining recoverable reserves of conventional oil, in gigatonnes; distribution of estimated values (from Delphi poll)

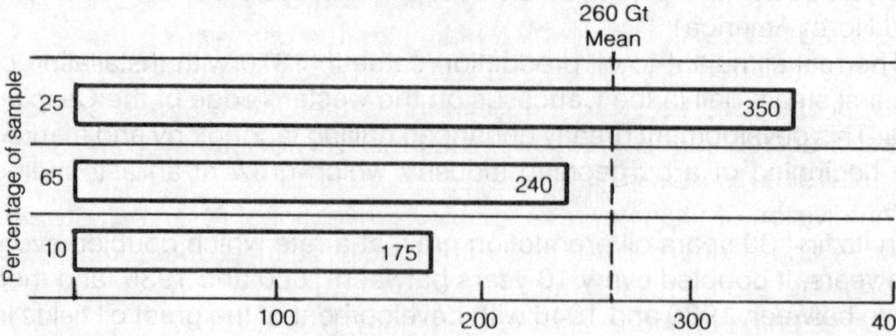

Chart 7

"conventional petroleum" was defined as that crude oil together with natural gas liquids which is economically recoverable using current technology under currently attractive economic conditions. In practice, it is the petroleum now being recovered onshore and offshore at a water depth of less than 650 feet.

As might be expected, the major portion of the ultimate conventional resources—42%—is to be found in the Middle East and North Africa (*Table 5*). The USSR, Eastern Europe, and China are next in ultimate reserves with about 24%. However, continued exploration in the USSR and recent announcements by the People's Republic of China could add significantly to total world resources and put these countries in a position rivaling the resources of the Middle East.

Table 5 Estimated ultimate recoverable conventional resources of oil

	Gigatonnes	%
North America	28.5	11
Western Europe	11.2	4
JANZ*	3.0	1
USSR, Eastern Europe, China	59.5	24
Middle East, north Africa	109.1	42
Central and southern Africa	11.3	4
East and south Asia	12.1	5
Latin America	22.9	9
Total	257.5	100

Delphi poll: mean values
*Japan, Australia, New Zealand

A 1948 estimate of the world's ultimate recoverable resources of conventional oil indicated a figure of 86 gigatonnes. Since that time the figure has grown steadily to 286 gigatonnes which, as the commission points out, reasonably agrees with its estimate of 260 gigatonnes. One of the difficulties encountered in evaluating ultimate recoverable resources, according to the commission, is the uncertainty in the recovery rate of crude oil from in-place resources. At present, the recovery rate may go as high as 25% with rates varying from field to field. While recovery rates are at present being improved by water injection and associated gas reinjection (which maintains the oil reservoir pressure), response to such secondary recovery is variable depending on the physical characteristics of the reservoirs. In the United States, secondary recovery techniques have increased recovery efficiency from 25% in the mid-1940s to nearly 30% today.

The division of ultimate, recoverable, conventional reserves between onshore and offshore locations is not readily defined, the commission reports, but opinions in the Delphi-poll converged on 60% of the resources being onshore, and 40% offshore. (*Chart 8*.)

With respect to the rate of future discoveries of petroleum resources,

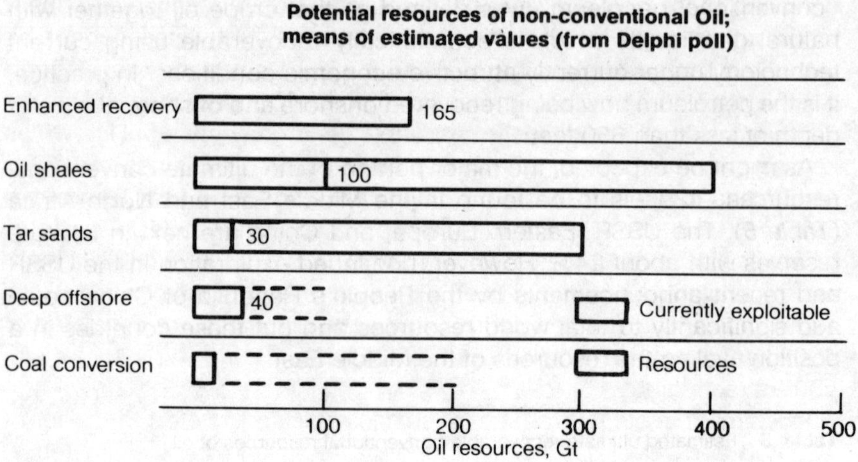

Chart 8

the generally expressed opinion is that petroleum reserves are approaching a finite limit, a limit which has been referred to repeatedly in recent years. Increased exploration costs, higher development costs, and limits to enhanced recovery techniques are indicative of this.

According to the commission, it is expected that future discovery rates will be approximately 4 gigatonnes a year by 1985 dropping to 3 gigatonnes a year by 2000. A widely held opinion is that the major part of oil resources will have been discovered by the year 2000. By 1985 it is expected that new fields will contribute 60% of the annual additions to reserves, but only 45% by 2000, the balance being made up by the re-evaluation of oil fields, enhanced recovery, and, possibly, the start of production of heavy oils.

A further concern is that discovery rates are considerably less than production. Between 1970 and 1975, for example, 1.5 gigatonnes of oil were discovered per year while the world was consuming 2.8 gigatonnes a year.

Unconventional Petroleum Resources

As a result of the concern over future availability of oil from conventional resources, much attention has been given to unconventional petroleum—a term applied to those oil deposits which will require exploration and exploitation technology not yet fully developed, so that oil production from them appears marginal today.

These unconventional resources generally include oil shale, tar sands, heavy-oil deposits, enhanced recovery from old fields, deep offshore deposits, polar deposits, synthetics from coal, and synthetics from biomass. Resource estimates, the commission adds, are very difficult and at present economic conditions do not support extensive exploration.

Among these non-conventional resources, potential reserves appear to be rather large in oil shales and tar sands, but these cannot be recovered by conventional production methods since they range from very viscous to semi-solid materials. (Oil is produced from shales by distilling crude oil and gases from a wax-like, organic material known as kerogen; in the case of tar sands, heavy bitumen is extracted.) Tar sands, and to a lesser extent oil shales, are also found near the earth's surface and can be mined by open-pit methods. To date, however—since resources of conventional, less expensive oil have been readily available—both have been exploited on only a small scale.

Oil Shales. Current petroleum reserves in oil shales are estimated at 400 gigatonnes—of which about 30 gigatonnes are exploitable with current technology.

Investment costs to produce oil shales are said to be on the order of $20,000 per barrel-day, with production technical costs estimated at between $20 and $25 per barrel (in 1976 dollars).

According to the commission, production of oil from shale should be underway by 1985 with production possibly up to two million barrels a day by 2000. However, ecological challenges will be considerable (disposing of tailings, avoiding contaminating water), initial investment in demonstration plants will be significant, and private investors will possibly require financial support from governments.

Tar Sands and Heavy Oils. While indicating that estimates of oil resources from tar sands and heavy oils are very provisional, the Conservation Commission projects reserves on the order of 330 gigatonnes—of which 15-30 gigatonnes are in surface deposits.

With tar sands, initial production of crude oil on a commercial scale began in Athabasca, Canada, in 1967. Today, tar sands are also being exploited in Albania, China, the USSR, Romania, Trinidad, and Venezuela. Canadian estimates show in-place worldwide resources at just over 327 gigatonnes, with 90% located in Canada, Venezuela, and the USSR.

According to the commission, a concept which holds promise of exploiting deeper oil sands deposits is on-site recovery. The key to realizing commercial production from these sands (sands that are too deeply buried to be economically mined), will be development of techniques for reducing bitumen viscosity to such a level that it can readily move through the permeable oil sands to the producing wells. Two basic techniques for heating these deep, underground oil reservoirs have been tested in Canada and other parts of the world: steam injection, and on-site combustion. While it is estimated that more than 30 gigatonnes of crude oil may ultimately be recovered from deep deposits, field pilot tests—several of which are currently underway in Canada—will be required to evaluate production methods.

Enhanced Recovery. Primary recovery from oil fields using present techniques is capable of extracting only up to 25% of the oil in place.

Secondary, or improved recovery by injected water and gas pumped back into the reservoir may lead to recovery of an additional 10% of the oil in place. The reason this greater part of the resource is left untouched is that the oil is trapped in the interstices of the reservoir rock formation and difficult to extract and raise to the surface as pressure falls.

To improve these percentages, and as one result of the dramatic increase in oil price in 1973, investigation has increased into "enhanced-recovery methods." Enhanced recovery involves the injection of heat, miscible hydrocarbons, and chemically modified water (surfactants, polymers) into the reservoir, and can increase oil recovery in a field up to possibly 45-60%. Inasmuch as this technology can be applied to known deposits, risks of exploration are not entailed. In addition, it can increase production considerably since such recovery processes can be applied to many currently active fields, to some depleted fields, and to future discoveries.

The amounts of oil that could be produced using enhanced recovery techniques are theoretically very high. For example, assuming 650 gigatonnes of in-place resources, an increase of the recovery factor to 50% would move recoverable reserves to 325 gigatonnes versus the 160 gigatonnes capable of being produced with today's recovery techniques.

In the United States it is estimated that about 7 gigatonnes of oil can be recovered under any set of economic conditions using current and foreseeable enhanced-recovery technology. In addition, it is generally felt that the responsiveness of enhanced-recovery potential to increases in the real price of oil would drop off about $22 a barrel.

Deep Offshore and Polar Deposits. The seas cover approximately 70% of the earth's surface with about 80% in the realm of deep ocean (depths exceeding 650 feet). On the basis of plate tectonics studies, most geologists feel that this deep ocean area contains little in the way of oil resources. Recent deep-ocean-bottom core samples further indicate that the sedimentary layer is very thin and covers a basaltic crust where the presence of oil is highly unlikely.

Despite the great uncertainty with regard to petroleum resources in the deep ocean area, some authorities have optimistically ventured to estimate resources there as equal to the present world recoverable oil resources of 230 gigatonnes. Other authorities estimate that possibly as little as 16% of the world's petroleum resources might be found there. This discrepancy can only be resolved by drilling, for which—although technology is now being developed—costs are not well defined.

As a result of these uncertainties, interest is concentrating on the continental margins of the oceans which constitute about 20% of the ocean's surface. The submerged sedimentary area of the continental margins is on the same order of magnitude as all onshore sedimentary areas.

As regards polar zones, experience to be gained in Alaska, northern

Canada, the Arctic islands, and northern Siberia will be of considerable value. Obstacles to developing these resources (once discovered) are many, including the protection of wellheads in the sea floor, submarine piping under ice packs, and drilling and production in offshore polar regions.

Synthetic Fuels from Coal and Biomass. Because of the geographic dispersion of coal deposits, and coal's abundance in many of the industrialized countries, the manufacture of synthetic oil from coal is attractive. In addition, coal resources are very large when compared to oil resources.

The production of synthetic fuels from coal to date has focused on converting coal to liquid and gaseous fuels for use in electric generation plants that presently burn either oil or gas. Despite the complex problems involved in conversion, it is estimated that by the 1990s commercial coal-conversion plants may be processing thousands of tons of coal a day and producing as much as three barrels of synthetic oil per ton. In the same period, it is estimated that coal gasification plants could be producing about 1940 to 3000 billion cubic feet of synthetic natural gas using 180 million tonnes of coal per year in the process.

The production of synthetic fuels through biomass conversion—the transformation of cellulosic materials, urban waste, and seaweed by hydrogenation or pyrolysis to synthetic fuels—is now at the demonstration plant level of technology. Inasmuch as biomass conversion is an indirect form of solar energy, it is therefore a renewable resource. According to the commission, at the onset of production this will be a high-cost energy resource.

Future Production of Conventional Oil

In its work the commission also projected possible, maximum, annual conventional oil production rates. This work takes into account rates of future oil discovery and a range of feasible oil depletion rates. Assuming a medium depletion rate, the results show a production ceiling of 5 gigatonnes a year reached by about 1990, with production then falling off to about 4 gigatonnes in 2005, and about 2.3 gigatonnes by 2020.

According to the commission's assessment, there probably will be sufficient oil from all sources to meet world demand for oil at the turn of the century only if oil is reserved for specific uses such as transportation and the chemical industries. The use of petroleum for heating and power generation, it predicts, will be greatly limited in the future having been replaced by coal and nuclear energy.

Additionally, to achieve production rates of 4 to 5 gigatonnes a year of oil in the decade 1990 to 2000, and to maintain a reasonable production rate for several decades thereafter, a timely and very large investment of capital in production, shipping, and marine facilities will be required. Furthermore, there are requisite technical, economic, and political conditions which must prevail if incentives to produce and distribute petroleum on a worldwide basis are to to develop.

Exploration and production technologies required to develop unconventional oil reserves will require extensive research and development, the commission says. Outside the centrally-planned economies, arrangements between governments and private companies must allow the companies sufficient incentives to supply the anticipated, very large capital requirements for research and technological developments needed to develop unconventional sources of oil. In addition, a skilled, technical labor force will be needed to support the new technology for operations in the deep offshore areas and to apply enhanced-recovery techniques to existing oil fields. Failure to provide funds, to invest in the required research and development, and to train and retrain the necessary skilled manpower, will create serious restraints to developing petroleum reserves from unconventional resources, it is said.

The commission also stresses that maximum oil production in the next few decades will be dependent upon the rate of discovery of new resources and the development of effective, enhanced-recovery systems. The maximum production forecast depends upon heavy expenditure of capital for exploration in unexplored areas, and in recovery enhancement in existing oil fields throughout the world. In particular, the commission says, exploration and development efforts must be made within geologically attractive areas of the less-developed nations whether these nations are currently producing oil or not.

Because of the shift of monetary resources from industrial nations to the oil-producing nations, the commission argues that oil-producing less-developed nations must commit funds to develop their own resources and, in addition, provide the capital to finance investment for petroleum exploration and development in the currently non-producing less-developed nations. International agreements may be required to ensure that adequate capital is available to finance the exploration and development of petroleum resources within these nations.

Another point made by the commission is that changes in current petroleum pricing practices and tax policies will be required to establish or re-establish sufficient incentives to provide the technology, manpower, and capital to continue the search for adequate supplies of oil. Changes of comparable magnitude will also be required in the fiscal and regulatory relationships between governments and companies, in both the private and public sectors, if maximum oil resource development is to materialize. This is true in the oil-exporting nations as well as in the industrialized nations (apart from the centrally-planned economies).

While the limits imposed by nature and technology on the development of resources are becoming increasingly significant, problems and challenges in the field of international politics are of even greater importance, according to the report. Not only is the decision to install oil production facilities one of political importance to many oil-producing nations, but also the further decision to operate these facilities at high levels of production is becoming a very serious political issue, since production means the export of a valuable resource patrimony. One of the principal challenges in the commission's eyes is to convince world policy makers that decisions must be made now to commit the required investment in capital and manpower to ensure an adequate future supply of petroleum. Failure to make these hard decisions, it is warned, will result in serious energy deficiencies with grave social consequences in the last decade of this century and beyond.

Natural Gas

Natural gas has become a much sought after fuel, providing over 17% of the world's primary energy requirements in 1978.

Proved reserves of the resource, having doubled over the last ten years, are equivalent to a life of over 35 years at predicted consumption levels. When the estimated remaining conventional recoverable natural gas resources are added to this total, indications are that gas supplies should be available well into the twenty-first century.

In addition, augmentation of these supplies by synthetic natural gas produced from other primary energy resources (such as shales, coal-bed degasification, and geopressured formations), although expensive to develop, should add significantly to total supply after 2020.

In view of both the potential market for gas, as well as the great

World and US production of natural gas

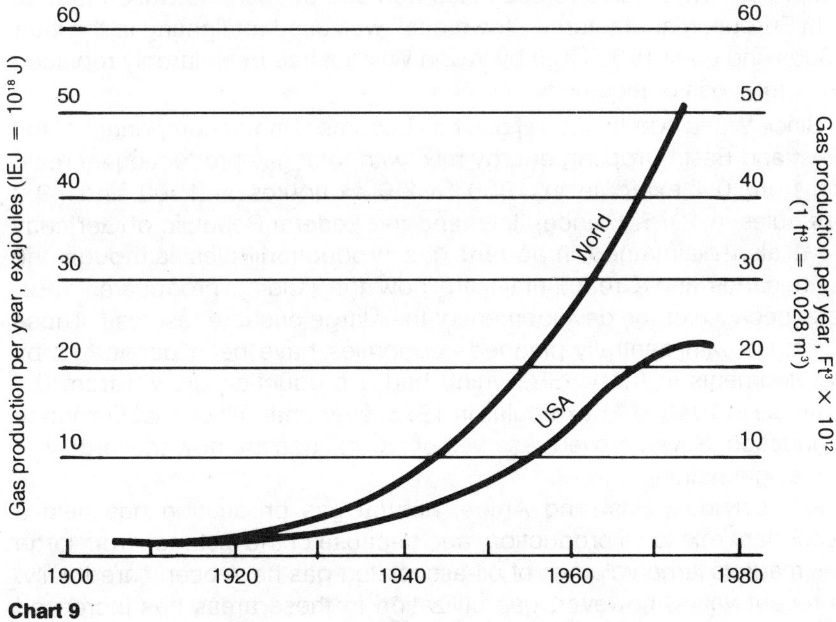

Chart 9

operational advantages and large capital plant existing in the gas industry's transmission and distribution systems, the industry is undertaking significant research and development into new processes for the production of synthetic natural gas and non-conventional gas. Although the production cost of gas will increase, and huge projects will have to be planned, designed, and constructed before projected supply can become a reality, the Conservation Commission projects that gas will remain competitive with other premium fuels and retain an important place in the market well into the middle of the next century. (*Chart 9.*)

The use of natural gas for fuel has a long history, but only in the past 50 years has it become one of the most important primary sources of energy for heating and industrial uses.

While the earliest use of natural gas cannot be pinpointed exactly, as early as 1000 B.C. the Chinese are said to have used a well system of bamboo tubes extracting gas from the earth for use as fuel in a brine-evaporating salt plant. Gas wells may also have been operating as early as 615 A.D. in Japan, while India, Persia, and a temple near Baku on the Caspian Sea were early users of natural gas. In Great Britain, Thomas Shirley in the mid-1600s discovered a spring "which roiled up near a coal mine and could be ignited with a candle."

The birthplace of the national gas industry in the United States is identified as Fredonia, New York, where in 1821 natural gas was first discovered. By 1891 a 120-mile gas pipeline was laid from northern Indiana to Chicago, Illinois, marking the beginning of high pressure, long-distance, natural-gas pipelining. A 1,000-mile, 24-inch pipeline was constructed in 1929 from the Texas panhandle to Chicago, and by 1948 a pipeline network of 346,000 miles was operating in the United States.

In Europe, manufactured "town gas" was used for lighting in the later 1800s and early 1900s, but by World War II it had been largely replaced by natural gas or electricity.

Since World War II, natural gas has become a major component of the West and East European energy mix, with total gas production increasing from 0.4 exajoule in 1950 to 2.6 exajoules in 1960 and 19.3 exajoules in 1975. France, Italy, and the Federal Republic of Germany have also maintained important gas production roles, although the Netherlands and Great Britain are now the leading producers in Europe because of the development of their large offshore reserves. Those countries with centrally-planned economies have been dominated by developments in the USSR, which had a production growth from 0.1 exajoule in 1945 to 11 exajoules in 1975, over one-half of total European production. Soviet proved reserves of natural gas are now the largest of any single nation.

In the Middle East and Africa, natural gas production has held a secondary role to oil production, and because of the distance from large gas markets large volumes of oil-associated gas have been flared daily. In recent years, however, gas utilization in these areas has increased from 0.25 exajoule to 2.6 exajoules (1965 to 1975), and construction of large and costly new liquefied natural gas and liquified petroleum gas plants indicates that future gas utilization will significantly increase.

While in some nations a substantial portion of oil-associated gas is still being flared, in many industrial regions there has been rapid development of gas gathering and transport systems. Indeed, according to the commission, the 1970s are witnessing the design, construction, and development of extensive new multi-billion-dollar systems of this type, both onshore and offshore, including liquid natural gas facilities, to utilize available gas and develop natural gas resources.

As regards liquid natural gas, it was initially produced on a commercial basis in 1940 in the United States with the first seaborne shipment coming 19 years later. At present there are 30 liquid natural gas projects under construction or planned, involving almost 11 exajoules per year. Before 1990 over $30 billion will be invested in liquid natural gas systems, and in the next several decades it is expected that liquid natural gas will become a major method for inter-regional gas transportation.

Conventional Natural Gas

Conventional natural gas resources include the total of proved reserves and undiscovered natural gas resources estimated to be recoverable with normal improvements in exploration and production technology. Such estimates have been done by a number of groups, with a compilation showing a general average of 2,500 exajoules for proved reserves, and approximately 8,150 exajoules for undiscovered recoverable resources. Total estimated conventional gas resources, therefore, stand at approximately 10,500 exajoules. (Adding estimated world use of natural gas up to 1975 to this total, it is projected that

World proved reserves and remaining undiscovered natural gas by world groups, CPE, countries with centrally planned economies; OECD, countries belonging to the Organization for Economic Cooperation and Development

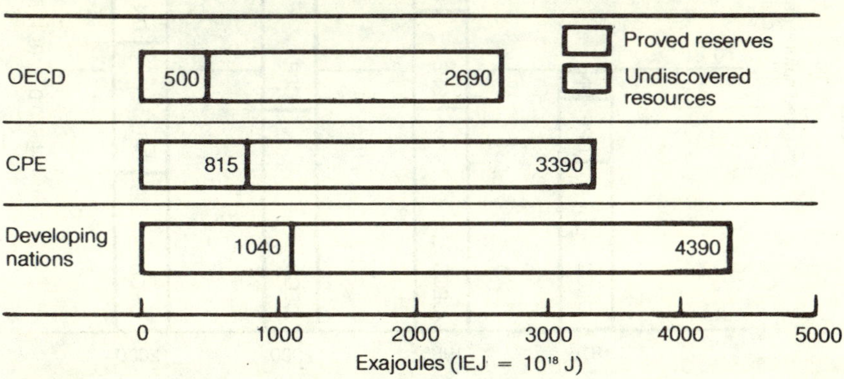

Chart 10

original total recoverable conventional resources of gas in the earth's crust were about 11,500 exajoules.)

According to the commission, developed nations account for about 25% of total world resources of natural gas, centrally-planned nations approximately 32%, and less-developed nations some 42%.

Future production of natural gas from conventional resources, the commission reports, is feasible at levels well above today's 50 exajoule a year output for at least a period of another 50 years. Assuming natural gas prices at a rate equivalent to $20 a barrel of oil (in 1974 dollars), the commission projects that world production of conventional natural gas could rise to 77 exajoules by 1985 peaking at about 143 exajoules by 2000. At this rate of increase, 4.4% a year, it is projected that production would reach a maximum of about 140 exajoules shortly after the year 2000, declining to 125 exajoules by the year 2020 (at which time about 50% of the presently estimated conventional gas resources would have been produced). (*Chart 11.)*

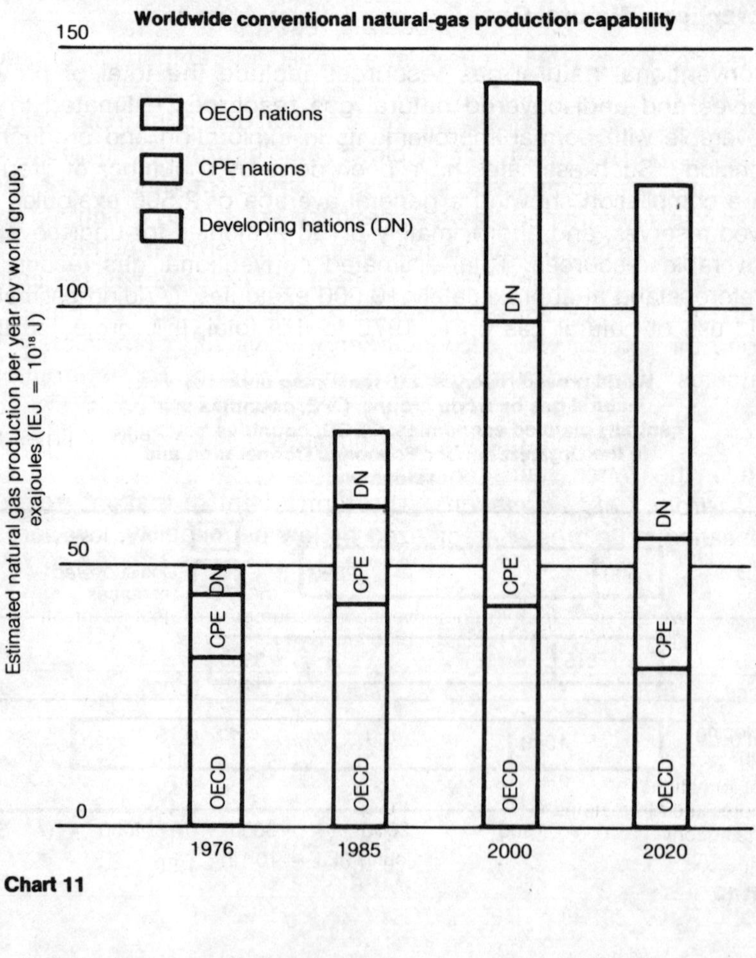

Chart 11

The commission points to the OPEC group of nations and the USSR as key areas of the world where substantial potential exists for greatly increasing production of conventional natural gas over the next decade, and to North America and Western Europe as regions in which it expects production to peak before 2000. Several areas of the world where exploitation has hardly begun could still be increasing conventional gas production by 2020, it notes.

Non-Conventional Natural Gas

Non-conventional gas resources addressed by the commission include methane from coal beds, gas from shales, gas from tight formations, geopressured gas, bioconversion, petroleum, and coal and peat conversion. (*Table 6.*)

Methane from Coal Beds. Methane occurs naturally in coal beds held there by absorption in the coal, in vertical joints and fractures in the beds, and in the interfaces with adjacent strata above and below the beds. Approximately 215 to 285 cubic feet of gas are associated with one short ton of coal. Only moderate recovery of methane from the world's coal beds would represent a significant contribution to natural gas supply.

Gas from Shales. Shales are rock formations containing kerogen which is believed to consist of biotic remains trapped in prehistoric sediments. The compaction process of shale formation traps these remains in fine-grained, low-permeability structures which serve as reservoirs. Gas and oil can be recovered from kerogen by heating, the gas representing about 10% by weight of the products generated. Among the world's shale formations, those in the United States are among the largest with a gross energy value of approximately 500 exajoules. (Much of this gas, however, cannot be economically produced with existing technology.) On a worldwide basis, the commission estimates that gas trapped in shales may be several times that found in the United States.

Gas from Tight Formations. The term "tight formation" refers to gasbearing structures characterized by low permeability, low porosity,

Table 6 World gas from non-conventional sources (exajoules or $ft^3 \times 10^{12}$) ($1\ ft^3 = 0.028\ m^3$). Production potential

Source	*1985*	*2000*	*2020*
Coal beds	0.01	0.15	2.7
Shale	0.01	0.10	2.6
Tight formations	0.05	0.70	4.4
Geopressured formations	0.03	0.45	3.9
Biomass conversion	Negl.	0.10	0.9
Total	0.10	1.50	14.5

and lack of structural continuity caused by complex cross-bedding and variations in clay and sand content. Tight formations characterize much of the geology of the Rocky Mountain areas and presumably exist in other regions of the world as well. The estimated resource potential of these formations in the United States is substantially over 650 exajoules, and considerably more for the world as a whole.

Geopressured Gas. Geopressured gas in the earth's crust is potentially the unconventional gas resource of greatest significance. In one zone alone—the Gulf of Mexico—its resource potential ranges up to 50,000 exajoules. These geopressured zones are found in some but not all of the world's coastal regions and are characterized by thick, sedimentary deposits containing water trapped at higher than normal pressures. This water is estimated to contain quantities of dissolved methane and to be hot enough to provide a source of geothermal energy. In addition to the Gulf of Mexico, the Siberian coastal basins, the Indo-China Sea, the Yellow Sea, the Sea of Japan, and the North Sea have structures which may contain geopressured gas.

There are real problems in developing this resource, however. The quantity of geopressured salt water required to deliver gas in commercial quantities will be so large that it raises doubts about whether onshore geopressured gas wells will be economically and environmentally acceptable. Furthermore, for offshore wells, there may be major environmental considerations centering on the large quantities of geopressured salt water which must return to either the sea or seabed formations, and the possibility of subsidence and fault activation.

Bioconversion. Methane gas can be produced through biomass conversion. The resource base for biomass methane includes a variety of animal and human wastes, and plant life, both onshore and in the sea. Biomass methane plants operate on the principle of anaerobic fermentation, with the gas thus produced used for domestic heating and cooking.

Research and development efforts in bioconversion are underway in many parts of the world. In industrialized nations such diverse potential sources as giant sea kelp, cellulosic materials, and city disposal wastes are subjects of current attention. In a number of less-developed nations, such as Nepal, Turkey, and India, biomass methane plants are already in commercial operation. It is conceivable that by 2020 biomass conversion to gas may be on the order of one to two exajoules per year. Inasmuch as biomass is a renewable resource depending on solar energy, it is therefore inexhaustible.

Petroleum, Coal and Peat Conversion. According to the commission, substantial amounts of gas should be produced by the end of this century by conversion from petroleum and coal resources. Two of the best-known processes are those involving the conversion of coal to gas of high calorific value, and the production of synthetic natural gas from liquefied petroleum gases, naphtha, and crude oil. Thirteen synthetic natural gas plants are now operating in the United States using liquid petroleum gases and naphtha.

Coal conversion technologies currently being pursued are systems for producing gases of high and low calorific value, coal liquefaction, and solvent-refined coal liquefaction and solvent refined coal-product production. These systems are costly at present, but price—on the order of $25 a barrel of oil equivalent—is approaching the marginal price of gas. On-site coal gasification which would provide energy recovery from coal seams not accessible with conventional mining techniques is another conversion process being developed by industrialized nations.

Development of commercial processes for peat gasification would add significantly to the world's non-conventional gas resources. A fuel resource found in many regions throughout the world and long used as a heating fuel, peat has in recent decades become an important source of fuel for industry and electricity generation in several European countries.

Natural Gas Export Trade

Summarizing its study of natural gas resources, the commission reports that it anticipates significantly increased international trade. One reason for this is that while gas demand is expected to increase at a lesser rate than production, it is projected that demand will exceed regional production capabilities in developed countries by 2000, and in centrally-planned countries by 2020. (*Table 7.*) By 1985 the commission

Table 7 World gas export trade projections (exajoules or $ft^3 \times 10^{12}$)($1\ ft^3 = 0.028\ m^3$)

	1985		*2000*		*2020*	
	Import	*Export*	*Import*	*Export*	*Import*	*Export*
OECD						
North America	0.9		2.6		18.3	
Western Europe	1.7		2.65		10.8	
JANZ	0.9		2.5		0.9	
Sub-total	3.5		7.75		30.0	
CPE						
USSR/Eastern Europe		1.0		1.0	10.5	
China/Asia					4.9	
Sub-total		1.0		1.0	15.4	
Developing nations						
OPEC		2.5		6.75		45.4
Non-OPEC						
Sub-total		2.5		6.75		45.4
Total export trade		3.5		7.75		45.4

envisions inter-regional export trade of 3.5 exajoules, ten times greater than at present. By the year 2000, the commission says, total international trade in natural gas—including 7.75 exajoules of inter-regional gas—will reach the level of ten exajoules a year and the developed countries will be importing at least 8% of their requirements. At the same time, the USSR, China, and the OPEC (Organization of Petroleum Exporting Countries) countries will have potential production capabilities greatly exceeding both their demands and those of importing nations.

Looking ahead to 2020, the commission reports that only the OPEC nations will have production potential exceeding indigenous demand. Assuming international transortation and distribution facilities exist, it adds, OPEC nations may be capable of meeting the requirements of other regions up to 2020.

Coal

Records of coal production on a worldwide scale do not date much further back than the later decades of the nineteenth century. However, in Great Britain the mining of coal began in the twelfth century, and by the year 1750 production had reached seven million tonnes. World production of coal is estimated to have increased to 134 million tonnes by 1860.

Today, production of coal stands at approximately 2.6 billion tonnes of coal equivalent. Beginning with industrialization in the mid-1800s, coal production doubled about every 16 years until World War I. In the period 1910 to 1950, production grew at a much less rapid rate owing principally to displacement by oil and natural gas in most sectors of the economy. Beginning in the late 1950s, however, the rate of growth in coal production again began to increase with total production in 1975 reaching 2.6 billion tonnes. While growth from 1860 to 1910 was approximately 4.2% a year, in the period 1910 to 1950 it decreased to less than 1.0% a year. For the past two decades (1955 to 1975) the rate has approximated 2.3%.

In the past the coal industry has met the world's demand for coal with timely installation of mining and transportation facilities, and recent studies continue to support expectations that coal resources will be sufficient to meet the world's energy requirements for centuries to come.

However, the challenge of the future with coal is not one of resources, it is one of production and distribution. If coal is to play its role as a major source of primary energy, a greatly accelerated program of mine development and transport construction will be required. A real concern in this regard is that men, money and materials may not be committed in time to meet the exponential growth in new demand.

According to commission estimates, annual production of coal will reach 8.8 billion tonnes for the year 2020. While a growth rate of 2.7% a year in production will accomplish this, some authorities are projecting rates as high as 4.3% a year to 2020.

Table 8 Production and export data for the main coal-producing countries (tons × 10^6)

	1975			*1985*			*2000*			*2020*		
	Production	*Export*	*Export (%)*	*Production*	*Export*	*Export (%)*	*Production*	*Export*	*Export (%)*	*Production*	Export	*Export (%)*
Australia	69	29	42	150	60	40	300	180	60	400	240	60
Canada	23	12	52	35	15	43	115	40	35	200	65	32
China (People's Rep.)	349	3	1	725	7	1	1 200	30	2	1 800	50	31
Germany (Fed. Rep.)	126	23	18	129	25	19	145	30	21	155	30	19
Great Britian	129	2	2	137	10	7	173	10	6	200	10	5
India	73			135	7	6	235	13	7	500	32	6
Japan	19			20			20			20		
Poland	181	39	21	258	45	17	300	50	17	320	50	16
South Africa	69	3	4	119	23	19	233	55	24	300	60	20
USA	581	60	10	1842	68	8	1 340	90	7	2 400	145	6
USSR	614	26	4	851	37	4	1 100	50	51	1 800	60	3
Other countries	360	2	1	483	6	1	619	34	6	751	46	6
Total	2 593	199	7.7	3 884	303	7.8	5 780	582	10.1	8 846	788	8.9

Coal Resources and Reserves

In assessing world supplies of coal, the commission is specific with its definitions of "resources" and "reserves." Resources are defined as those deposits that at some time in the future may acquire an economic value for mankind, while reserves are defined as those resources which are exploitable under current technical and economic conditions.

Using these definitions, it is projected that total world hard- and brown-coal resources today represent 10,125 billion tonnes of coal equivalent. Of this, reserves conservatively total 637 billion tonnes of coal equivalent. Of this, reserves conservatively total 637 billion tonnes of coal equivalent—or 6.3% of resources. (Differentiation between brown For the commission's study, the demarcation was established at 10,250 BTUs—British Thermal Units—a pound. Bituminous and anthracite coal fall into the category of hard coal, and lignite and sub-bituminous coal fall into the category of brown coal.)

The major portions of these coal resources and reserves are found in the northern hemisphere, principally in the USSR, China, and the United States: 86% of the hard-coal resources are found in these countries, and 94% of the brown-coal resources. The corresponding figures for technically—and economically—recoverable reserves are 60% for hard coal and 64% for brown coal.

As an addendum to these figures, the commission points out that its estimates for both resources and reserves may be too low. For resources this is because of the minimal exploration effort to date in many areas of the world (especially the southern hemisphere). In addition, in major coal-producing countries, resource data reported appear to be conservative. As regards reserves, low estimates may be due to mining techniques which exploit only part of the in-place reserves; to mine planning which omits many of the deeper reserves; and to practices that prohibit mining beneath populated areas.

It might be expected that as recoverable reserves of coal are increased the ultimate production rate might increase as well with higher production extending over a longer period. If this thinking is applied to recoverable reserves of 600 billion tonnes of coal equivalent, the production projected for 1985 and 2000 can be met satisfactorily; however, production at a rate of 8.8 billion tonnes of coal equivalent cannot be met in 2020. On the other hand, if recoverable reserves were to double from 600 to 1,200 billion tonnes of coal equivalent or more, a production rate of 8.8 billion tonnes of coal equivalent or more could be met quite readily.

In its report the commission says it is possible to conceive that the economics of coal recovery may change so that 20% to 25% of the geological resources will be in the form of recoverable reserves by 2020.

If this were to develop, coal production could achieve levels significantly beyond 8.8 billion tonnes of coal equivalent by 2020.

Indeed, a review of production rates shown by the larger coal-producing countries has led the commission to conclude that production can increase more rapidly than indicated. However, if this is to be achieved a demand must develop so that there is an acceptable market price and potential consumers will commit themselves to long-term contracts that offer an appropriate rate of return on the investments required.

One possible production projection made for the United States shows a growth rate of 7.2% in the decade 1975 to 1985—a rate based on past experience in the U.S. and the fact the easier reserves will be mined in this earlier decade. In future years, however, mining will be more difficult, resulting in a progressively decreased rate: 3.8% from 1985 to 2000, and 3.2% from 2000 to 2020. However, average production growth with the proposed trend does increase the country's rate from 3.2% to 4.3% over the entire period 1975 to 2020. If this same rate were used for Australia, China, South Africa, and the USSR, it would increase the combined production of these nations from 7.9 billion to 12.3 billion tonnes of coal equivalent by 2020, and the total for all the nations would then be 13.1 billion tonnes of coal equivalent.

Conditions Affecting Coal Availability

In its analysis of coal resources the commission lays important emphasis on conditions affecting future availability and use. Included here are discussions of future mining centers, mine development lead times, infrastructure requirements, capital and labor requirements, and environmental impacts.

Future coal mining centers, USA

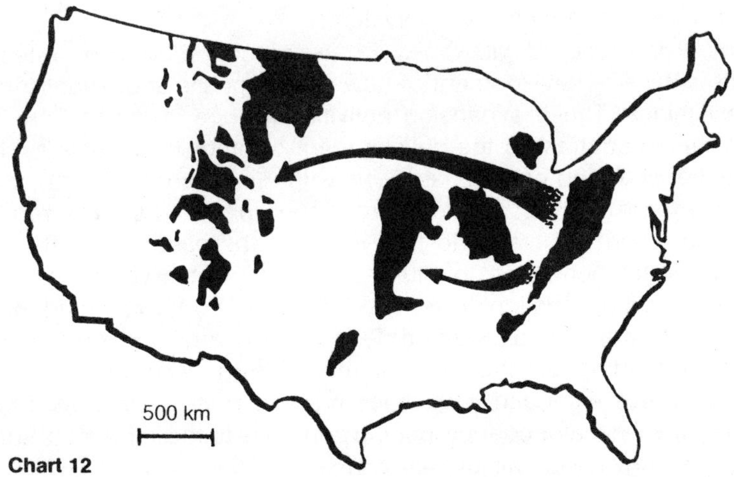

Chart 12

Future coal mining centers, USSR

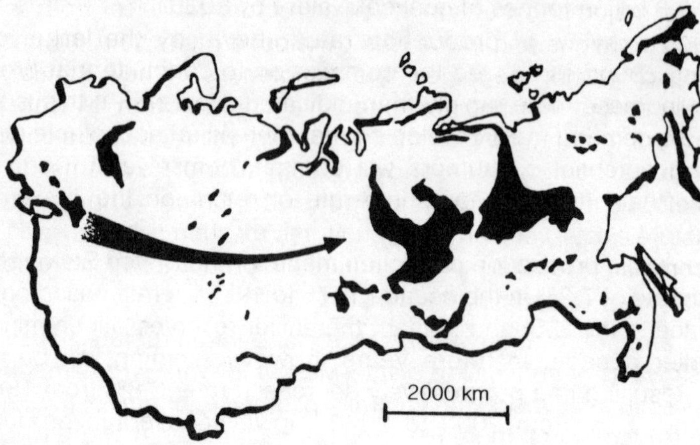

Chart 13

Future Mining Centers (*Charts 12 and 13.*) While the major coal-producing countries have increased exploration activities to discover and develop new deposits, the Conservation Commission contends that a significant increase in production in many of these countries is possible only by establishing new principal mining centers in areas which are less developed industrially. For example, although the U.S. coal production base will remain in the eastern portion of the country for some time, a transfer of coal mining operations will eventually take place from the eastern to midwestern and western states. In the USSR the deposits in the European sector have a limited lifetime, calling for a gradual transfer of principal operations to coal deposits east of the Urals and Siberia over the next few decades. The opening of these mines at locations distant from industrial energy users will require the construction of sizeable energy transport systems.

Mine Development Lead Times. Projected coal production rates will require extensive development of new coal deposits and establishment of new mines. These expansion activities require comparatively long lead times in contrast to the development of other fossil fuels such as conventional oil and natural gas. Even with the existence of reasonably adequate infrastructure, development of open-pit surface mines requires a minimum period of three to five years. In the case of underground mining development, it is now taking a period of ten years from mine opening to full production. When the time required for securing mining permits is added, a lead time of between six and twelve years—depending on the type of mine being developed—is more realistic. These extended lead times make it quite clear that a rapid response by the major coal-producing countries to the immediate supply of world energy needs will be difficult.

Infrastructure Requirements. Worldwide projected coal production will require major construction of infrastructure facilities, both in the industrialized nations and in the less-developed nations. In the case of major coal-producing countries, the remoteness of new mines from industrial centers will require extensive construction of transport facilities. Furthermore, the new mining centers will require the development of large communities and industry support systems. In the less-developed nations, such as India and China, where minimal infrastructure exists, very large expenditures for such new transport and mining community facilities will be required to produce and move coal to domestic energy-consuming centers.

Capital and Labor Requirements. Mine and infrastructure needs and the construction of conversion plants will call for very large amounts of capital investment. Outside the centrally-planned economies, this capital will be provided only if there are credible guarantees that the invested funds will be adequately serviced, a concern which may hinder the timely supply of coal by the investor-owned coal industry. However, as demand develops and supply contracts are secured, it is believed that the financial market will be capable of providing the required capital.

Regarding labor requirements, and even in the face of future automation, a greatly expanded labor force will be required to support projected growth in coal production. Here again significant lead time will be needed to recruit and train qualified mining personnel. In many cases, whole new towns must be created, meaning that labor must be willing to be mobile and adaptive to new environments. This factor of labor supply may well be a limiting force in coal industry growth in some of the large coal-producing countries.

Environmental Impacts. The trend in environmental standards in many coal-producing nations is to require not only environmentally compatible mine operating conditions but also the restoration of mined-out sites, both major concerns for mine operators. With coal users, environmental considerations for processing, converting, and utilizing coal have had an impact on demand. Concerns over sulphur dioxide and nitrogen oxide emissions, as examples, require extensive investment by users for appropriate control systems. In this latter regard, systems now in the demonstration-plant phase promise economic commercial application by 1985, thereby reducing the emissions deterrent to the utilization of coal.

Another environmental concern with coal use, however, is carbon dioxide, a product of combustion with all carbonaceous fuels. Concern here centers on increases in atmospheric carbon dioxide and the hypothesized "greenhouse effect," or warming trend, this might have on climate. To date there is no consensus on the impact of atmospheric carbon dioxide, however, and accumulating dust in the atmosphere is considered to have a cooling effect. Regardless, vigorous research efforts should be made in order to understand all future ramifications of carbon dioxide in the atmosphere.

World Trade in Coal

Most projections of future world energy supply show coal to be a major supply resource, providing by the year 2020 between one-quarter and one-half of the world's energy requirements. As a result, coal will most certainly become a major commodity in world trade as early as the year 2000. Indeed, on the basis of regional energy balance studies, it is conceivable that the role of coal may rival that of oil in world trade by 2010.

As a result, an important question is the extent to which major coal producing nations will make coal available for export. Currently, world trade volume of coal is around 200 million tonnes of coal equivalent annually, with transport divided evenly between overland and marine shipments. Expressed in terms of heating value, this is equivalent to about 5% of the current international trade in oil.

One distinction that exists between coal and oil as an international trading commodity is the typical country of origin for each fuel. Increasingly, oil is produced in developing nations where only a small fraction is used domestically, whereas coal is produced predominantly in industrialized nations which consume essentially their entire production. Therefore, an expansion in the export trade of coal will develop only if there is an attractive market. At the moment, it appears that in the absence of an acceptable market the main coal-producing countries have no plans for significantly increasing their world trade in coal. In comparison with the domestic consumption rates of the main coal-producing countries, the projected export quota remains relatively constant to the year 2020.

With projected coal demand in 2020 up to four times its present level, transport facilities will be crucial to both domestic and international supply. While this represents a significant challenge to the development of coal transport systems, there is a precedent in the experience of the iron ore industry over the past 25 years. One of the bottlenecks to future coal export trade, however, may be superport facilities for the seaborne coal.

Based on iron ore industry experience, it would appear that the development of world trade in coal need not be inhibited by seaborne transport. The iron ore experience corresponds to a 10% annual growth rate over the past 25 years. Coal being a commodity similar in nature to iron ore, it too should be able to achieve a comparable growth rate of seaborne transport.

Research and Development

If coal is to fulfill the energy supply role envisioned for it in the next forty years and beyond, significant improvements in mining technology must occur. This increased demand for coal, combined with pressures against manual labor in mining, is expected to result in increased research and development into highly automated mining processes.

Such research is focusing today on improving coal plow or ripper technology; on coal extraction by cutting and shearing; on feed systems, conveyors and their drive systems, and the rigging of mining and conveying facilities along wall faces; on the forces encountered with blockages to ensure safe structural design; and on metering output tonnage.

For working systems and methods for inclined seams the main research and development tasks now being addressed are improvement of efficient integrated equipment for conventional wall face mining; new support methods; and hydromechanized mining.

In the area of coal gasification, coal gas (produced by distillation when heating coal in a retort in the absence of air), water gas (produced by the reaction of steam with hot coal or coke), and producer gas (produced by the reaction of a mixture of steam and air on coal or coke) remain in use today as they have for over 100 years, with newer technology for production — as an example, substituting oxygen for air in the producer-gas process — being given increasing attention.

With higher fuel prices and the prospect of natural gas or oil supplies becoming inadequate, many major energy organizations have embarked on a new phase of developing coal gasification. Several possible variants of a coal-steam-air gasifier, as an example, are being investigated in the hope of finding a more efficient, economic continuous process which will advance the day when synthetic gas will supersede natural gas. This effort is receiving substantial support from governments, although there are as yet no major new demonstrations or commercial installations.

As regards coal liquefaction (i.e., the production of liquid hydrocarbons from coal), work continues in a number of areas. These include pyrolysis, where crude tar and light oils are produced as a by-product of the coking operation; indirect hydrogenation, where synthetic gas which can be converted to methanol and paraffinic hydrocarbons in the presence of a magnetic iron catalyst is produced from a reaction of steam and oxygen with coal; direct hydrogenation, where coal, usually slurried with a recycled oil, is reacted with hydrogen in the presence of a catalyst; solvent extraction, where fractions of coal are dissolved out with oils, asphalt, or solvent gas at elevated temperature and pressure; and auxiliary processes.

Nuclear Energy

Nuclear energy may be divided into energy from fission and energy from fusion.

Nuclear fission is the splitting apart of the nuclei of atoms into two fragments of approximately equal mass, accompanied by conversion of part of the mass into energy.

Nuclear fusion, on the other hand, is the recombination of the material in the nuclei of two atoms (called ions) of lightweight elements in such a way as to result in two particles of lower total mass and a corresponding release of energy. This reaction requires the intimate association of highly excited ions and, if the process is to result in more energy being released than used, this highly excited state must be maintained long enough for several reactions to occur.

Nuclear fission is by far the more dominant contributor to energy supply both presently and potentially to 2020. Indeed, no measurable amount of energy from fusion is expected by 2020.

According to the commission, nuclear power from fission is the outstanding example in the energy field of practical application flowing from newly discovered scientific fact.

Radioactivity, key to the structure of the atom, was discovered in 1896, and in less than 50 years the main constituents of the atom, their interrelationships, and various nuclear reactions had also been discovered. By 1939, the fission of uranium atoms by neutrons, first encountered in 1934, was understood, and in 1942 the first reactor based on this reaction was built in Chicago. By the early 1950s electricity was being generated by nuclear fission in both the U.S. and USSR.

By 1977, 153 nuclear-electric generating units in 17 countries produced 6% of the world's electrical energy, or about as much electrical energy as the world had been using when nuclear fission was discovered 40 years before. The first half of the 80-year period when the underlying science was developed, it is pointed out, was characterized by research in universities and in rather modest research institutes. The second half of the period, however, when application began, was marked by major research and development effort at special, large national laboratories and, to a lesser extent, at university and industrial laboratories.

The transition from scientific inquiry to large-scale application was accomplished in a remarkably short time and, according to the commission, affords some reassurance for the future about man's resourcefulness, given appropriate incentive and encouragement.

Energy From Fission

Uranium is the primary element used today in nuclear fission, but knowledge of world uranium resources is limited due to its short history as a source of energy and the wide fluctuations in demand for it. Indeed, in reporting resources of uranium the commission uses the term "world" to exclude the U.S.S.R., Eastern Europe, and China.

The mining of conventional uranium-bearing ores began on a small scale before the start of this century, and after 1940 developments in the new nuclear energy field stimulated the recovery of uranium from tailings discarded from earlier radium and vanadium operations in the United States, and permitted the resumption of recovery in Canada and Zaire. During the early 1950s, U.S. production from domestic sources of uranium gained momentum as a result of a guaranteed government market. At the same time, discoveries were made in Canada and production there increased with development of new domestic deposits. In 1952, uranium by-product recovery from gold-mining operations was started in South Africa.

By 1955, France—having succeeded earlier in developing important domestic vein-type uranium deposits—had expanded its exploration efforts to the sedimentary terrains of Africa with resulting discoveries in 1956 in Gabon. In the 1960s other important deposits were identified in Niger and the Central African Empire.

The less than 35-year history of uranium as an energy resource has been relatively short when compared with that of hydrocarbons, and it has been a past marked by peaks and valleys. The industry's initial, spectacular growth in the 1950s to satisfy demand for nuclear weapons was followed abruptly by a period of rapid decline during which the promise of a demand for nuclear power purposes was slow to materialize and the industry suffered for several years from surplus stocks, excess capacity, and low prices. However, following the oil crisis of 1973-74, its more promising future took form.

Uranium Resource Estimates

The Conservation Commission divides uranium resources into two classes based on assurance of existence—reasonably assured resources and estimated additional resources—with estimates of tonnages judged on the basis of what can be recoverable at costs up to $130 per kilogram of uranium. No attempt was made by the commission to quantify world resources exploitable at costs in excess of this figure, inasmuch as such estimates exist in only isolated cases.

Table 9 Estimated world uranium production capability to 1990 (tonnes)

World region	*1977*	*1980*	*1985*	*1990*
North America	20 800	31 550	48 500	58 250
Western Europe	2 560	3 850	5 940	6 700
Australia	430	530	11 030	20 030
Latin America	230	1 070	1 190	600
Middle East and north Africa		100	100	100
Africa south of Sahara	9 310	18 000	23 700	23 200
East Asia		30	. . .	. . .
South Asia	. . .	. . .	. . .	. . .
World total	33 330	55 130	91 260	108 880

( ) Nil; (. . .) no estimate available

Estimated reasonably assured resources of uranium today are put at 2,191,700 tonnes by the commission, with estimated additional resources placed at 2,176,200 tonnes. According to the commission, 38% of these reasonably assured resources are in North America. Africa south of the Sahara holds 25%, Western Europe 18%, Australia and Japan 14%, and other regions the remaining 5%.

Of the over 60% of the world's reasonably assured uranium resources in North America and south of the Sahara in Africa, most is said to be in sandstone-type deposits, quartz-pebble conglomerates, veins and related deposits. At the same time, some 77% of the total shown for Western Europe is contained in Swedish alum shales and, as a result, it is very difficult to predict if this is exploitable. The bulk of the reasonably assured resources occur in Australia, largely in vein and related deposits.

Beyond the reasonably assured resources of uranium, it is felt that some 79% of the world's estimated additional resources are located in North America—with slightly more than half in the U.S. The bulk of the total is also contained in sandstone-type deposits, quartz-pebble conglomerates, veins and related deposits. (The disproportionate size of the estimates for North America, the commission says, is largely a reflection of the amount of exploratory expenditures in that region as opposed to others.)

Of the uranium resources in both the reasonably-assured, and estimated-additional categories, the commission reports that over 70% is believed to be recoverable at costs up to $78 per kilogram of uranium. As a result, this total can be viewed as being of economic interest today.

Regarding undiscovered uranium resources, the United States is the only country which has published comprehensive, quantitative estimates. The most recent of these shows some 1.5 million tonnes of uranium in the U.S. In the case of most countries, the need to provide comprehensive assessments of these undiscovered resources has only recently become evident, and efforts have been severely hampered by the lack of basic geological, geochemical, and geophysical information, and the embryonic state of methodology.

In the absence of a comprehensive data base, several attempts have been made to estimate the world's undiscovered, recoverable uranium resources using mathematical/statistical models. Various estimates have been reported ranging from 80 to 280 million tonnes of uranium. Although the results of these efforts are encouraging, they are in no way conclusive and continue to be viewed with scepticism by most uranium resource exports. On the other side of the coin, however, an examination of known uranium resources in relation to the world's geological provinces indicates that it is not logical to assume that North America—with only 17% of the world's land mass (excluding that of the U.S.S.R., Eastern Europe, and China)—should prove to have 58% of total estimated, known resources.

Although uranium exploration activity has been reported in some 30 countries, only in the case of a small number of these could it be said that activity was substantial. Indeed, except perhaps for North America and Western Europe, no region of the world has received nearly the attention its uranium potential deserves. The overall problem, the commission concludes, is not likely one of existence of resources, but rather of discovering these resources and bringing them into production within the required time scale.

In addition to conventional resources of uranium, there are several types of operations from which uranium can be recovered as a by-product. For example, it can be a by-product of phosphoric acid production, of solutions generated in the leaching of certain copper ores, from monazite (which is produced in the heavy-mineral, beach-sand industry), and as a co-product of several other elements from Swedish alum shales. The possibilities of recovering uranium as a principal product from phosphate rock, marine black shales (e.g., Chattanooga shales), granites of above-average grade coals, lignites, and seawater have also been investigated.

Although many of these unconventional uranium resources are very large, their potential for contributing large increments of annual production is in most cases limited. The constraints vary from case to case and include such things as the lack of technology for recovery, the high cost of production even when the technology is known, dependence on the rate of production of the principal or co-product, the often vast scale of mining required for the very low-grade sources, and the environmental implications associated with exploration.

In view of these various and complex factors, the commission reports that unconventional resources should be expected to contribute only relatively small amounts to world uranium supply.

Thorium Resource Estimates

In addition to uranium, one other element occuring in nature displays the necessary properties and abundance to augment fissile nuclear energy resources: Thorium-232, which is nearly 100% of the naturally occuring element and will convert to fissile uranium 233 on capture of a neutron.

The world's reasonably assured resources of thorium, recoverable at costs up to $75 per kilogram of uranium, are currently estimated at some 630,000 tonnes, almost 50% of which is contained in monazite in heavy-mineral, beach-sand deposits in India. Much of the remainder is in similar beach-sand deposits primarily in Brazil, Australia, Malaysia, and the United States. These beach-sand deposits are currently being exploited for their heavy-mineral content (i.e., minerals of titanium, tin, and zirconium). Monazite is also recovered for its rare-earth content.

Present world production of thorium is on the order of 730 tonnes a year, all as a by-product of monazite. Should all of the thorium from monazite production be recovered, it is estimated that up to 1800 tonnes a year of thorium could be made available at current levels of production. If additional thorium were required it is estimated that world production levels could be expanded to at least 4200 tonnes a year by recovering thorium primarily as a by-product of uranium-producing operations in Canada, of copper mining in South Africa, and of niobium mining in Brazil. Still further increases in production could be achieved by exploiting thorium as a principal product from thorite vein deposits in the United States.

Uranium Production

World industry has to 1976 produced some 470,000 tonnes of uranium, approximately 68% of which has come from North America. The bulk of the remaining production has come from countries in Africa south of the Sahara. Present worldwide production is some 23,000 tonnes of uranium a year. The industry's peak year of production was 1959 when some 34,000 tonnes of uranium were produced.

Present production capability is estimated at about 33,300 tonnes a year of uranium (again with the vast majority taking place in North America and in Africa south of the Sahara). However, certain production facilities now being expanded, and new plants either under construction or planned, could—when completed—bring world production capability to about 55,000 tonnes a year of uranium by 1980. By 1990, known resources are believed capable of supporting a maximum level of production approaching 110,000 tonnes of uranium. According to the commission, levels of production much in excess of 110,000 tonnes a year of uranium are unlikely to be reached without identification of new sources of production.

Any projections of uranium production capability are subject to a number of constraints, however. In view of this, certain physical limitations on the rate at which particular deposits can be exploited, as well as the realities of declining grades and resource depletion, have been incorporated into the projections. Constraints have also been taken into account in a few cases where production rates in certain non-exporting countries are set by government policy, and where there are very clear limitations to availability.

Several other types of constraints on production are much more elusive, however, and may possibly have a negative effect on the projections.

Assumptions made about timing of uranium developments in Australia, for example, may prove to be optimistic; supplies of manpower and equipment may prove to be inadequate; increasing delays may be experienced due to more complex licensing and regulatory requirements; appropriate forward baseload sales contracts may be elusive; and adequate financing may not be forthcoming.

Pinpointing uranium production beyond 1990 also requires assumptions concerning the magnitude of the undiscovered resources which may be identified and developed by the turn of the century. If these undiscovered resources prove to be several times today's known resources, the commission believes it will be possible to support production levels considerably beyond the 110,000 tonnes estimated for 1990.

Nuclear Energy for Electrical Power Supply

While about 4% of the world's electrical requirements are provided by nuclear power plants today, by the year 2020 most demand projections anticipate that nuclear energy will provide the principal share of electrical power supply.

For the long term, and on a worldwide basis, it is projected that the fraction of electrical energy supplied by nuclear resources will tend toward 45% by the year 2000, and toward 60-65% by the year 2020. World total nuclear generating capacity implied by these percentages would be between 1300 and 1900 gigawatts (electric) in the year 2000, and between 3200 and 5500 gigawatts (electric) in the year 2020, depending on assumptions of economic growth for those periods.

Uranium Demand...and the Supply Challenge

In attempting to determine requirements for conventional uranium resources with respect to projected electrical power generation by nuclear power plants, the commission designed five nuclear scenarios to cover a wide range of nuclear growth possibilities. Of the five, two are said to illustrate the range in thinking on future use of reactor types.

One. This is the base case, using light-water reactors initially but installing commercial breeder reactors with a fuel doubling time of 24 years. The breeder reactors are installed beginning in 1993 in North America, in 1987 in Western Europe, in 1995 in the U.S.S.R., and in 2000 in Japan.

Table 10 Estimated annual world uranium demand (tonnes × 10^5)

Nuclear growth	*OECD countries*		*Centrally planned economies*		*Developing countries*	
scenario	*2000*	*2020*	*2000*	*2020*	*2000*	*2020*
1	1.5	1.7	0.7	1.7	0.4	1.8
2	1.8	4.2	0.8	2.8	0.4	1.8

Two. This is a "no reprocessing" scenario. In this case, only thermal converter reactors are installed in all regions. Light-water characteristics are used for these since this is the most prevalent commercial reactor in use today.

For these scenarios, annual world uranium demand is projected as follows (in 100,000's of tonnes):

Again using these scenarios, cumulative world uranium demand—cumulative from 1975 to the year indicated—is projected as follows (in millions of tonnes):

Table 11 Estimated cumulative world uranium demand* (tonnes × 10^6)

Nuclear growth scenario	*OECD countries*		*Centrally planned economies*		*Developing countries*	
	2000	*2020*	*2000*	*2020*	*2000*	*2020*
1	2.0	5.3	0.6	2.0	0.5	2.2
2	2.0	7.8	0.6	3.9	0.5	2.2

* Cumulative from 1975 to year indicated.

For the developed countries, only about a decade is required until annual uranium demand levels off if the breeder is introduced there (scenario one). However, the accelerating uranium requirement for countries in which it is assumed that no breeders are introduced reduces the impact of this effect on world requirements. Requirements continue to increase, but at a lower rate.

Total uranium commitments to the year 2020, including sufficient uranium to operate the nuclear plants installed by that time for the remainder of their assumed 30-year economic life, are shown as follows (in millions of tonnes):

Table 12 Estimated world uranium commitments to 2020 (tonnes × 10^6)

Nuclear growth scenario	*OECD countries*	*Centrally planned economies*	*Developing countries*
1	7.0	4.7	4.8
2	13.0	7.7	4.8

According to the commission, considerable effort will be required to make uranium available from known resources at the rates of demand assumed. Although the resources needed to support this capacity have largely been discovered, further effort will be required to identify and delineate resources in the estimated additional category. In addition, financial commitments have yet to be made for the projected additional capacity beyond about 1980, and the great bulk of production required to meet the growing supply/demand gap likely to begin in the late 1980s will need to come from new sources.

The nature of the world uranium supply challenge is emphasized by projections (using the assumptions described in scenario one) that production rates from known sources will decline after about 1990 owing to depletion of resources in some deposits and to the mining of lower-grade tonnages in others. According to the commission, and even assuming that the estimated maximum capability of 100,000 tonnes a year of uranium from known sources can be achieved, some 300,000 tonnes of annual production would have to be developed from new sources between 1990 and 2020.

Although a portion of this production gap may be met from currently known sub-economic sources, by lowering costs, and through advances in technology, the bulk of the additional requirement must be met by production from new sources. In this regard, the actual quantity of uranium resources that must be discovered will vary considerably, depending on the order in which deposits of different sizes and grades will be discovered and developed.

Assuming an average five-year lead time between discovery and initial production, the commission estimates that 329 new discoveries representing a mixture of six mine types would be required by the year 2015 to provide the levels of production needed by 2020. (*Table 13.*) Total "mine-life" reserves contained in these 329 new discoveries were estimated as some 9.5 million tonnes of uranium, and the total discovery cost (in 1976 dollars) was estimated at $50 billion. In addition, the total capital investment (in 1976 dollars) required to provide the new production facilities was estimated at between $30 and $40 billion.

Table 13 Illustrative discovery requirements to meet annual world uranium demand to 2020 under Scenario 1

	If all additional uranium products to the year 2020 is to be met from the finding and subsequent opening of a mix of new mines similar to:	*Number of discoveries required by the year 2015*
A.	Large-sized, low-grade, opencast mine; e.g. Rossing, Namibia	12
B.	Medium-sized, medium-grade, sandstone production centre; e.g. New Mexico, USA	78
C.	Large-sized, low-grade, underground mine; e.g. Denison, Canada	8
D.	Medium-sized, medium-grade, underground mine; e.g. Eldorado, Canada	60
E.	Large-sized, high-grade, opencast mine; e.g. Ranger, Australia	27
F.	Small-sized; medium-grade, sandstone production centre; e.g. Wyoming, USA	144
Total discoveries required to 2015		329*

* Some 50 of the total new mines are required due to depletion of deposits discovered in the early part of the period 1977 2015.

Fuel Utilization

Advances in the technology of reactor fuel cycles are required if the utilization of uranium resources is to be improved significantly.

Here, new enrichment techniques will probably be developed which may make it possible to slightly reduce losses in "tailings" from plants thereby decreasing uranium requirements per unit of enriched production. The maximum theoretical reduction in uranium demand would be 35% should it be possible to reduce losses to zero. Practically, with low-cost laser separation techniques, it is expected that concentration in tailings may be lowered to about 0.10% uranium-235. The further reduction to 35% could be more readily achieved by recycling uranium and plutonium in light-water reactors. Similarly, a reduction of about 50% could be obtained by recycling plutonium in heavy-water reactors. The most significant advances in the technology of nuclear power, however, will arise from the ability to use more advanced fuel cycles.

The use of the thorium fuel cycle in thermal reactors and the plutonium-uranium cycle in fast breeders can also reduce uranium requirements considerably, but the reduction obtained depends on the growth rate of the nuclear system. For this reason, nuclear fusion and the spallation reaction may become interesting in the future as a means of producing fissile material from fertile material. All advanced fuel cycles require fuel reprocessing, remote fuel fabrication techniques, and permanent waste storage in addition to the conventional "front end" of the fuel cycle. Partly because of the long lead times involved, fuel enrichment and fuel reprocessing could serve as bottlenecks in the commercial developement of nuclear power.

Reactors using advanced fuel cycles must also be developed. Except for the heavy-water reactor, which can use the thorium cycle with some modification, most of the candidates are in the prototype state. The most prominent of these prototypes are the high-temperature, gas-cooled reactor for the thorium cycle, and the liquid-metal, fast-breeder reactor.

Even here, however, fuel cycle characteristics must be improved if uranium and thorium resources are to be more fully utilized.

Advanced reactor fuel cycle technology would make possible the utilization of an impressive source of nuclear energy contained in the depleted uranium currently stored at enrichment plant sites in the United States. It is estimated that there is approximately 250,000 short tons of depleted uranium stored at these sites. If only 70% of this were used in fast-breeder reactors it would be possible to provide an energy-equivalent greater than the 178 billion tonnes of coal considered to be technically and economically recoverable from the U.S.'s entire coal resource. This depleted uranium requires no mining and is available for immediate shipment in a form convenient for processing into blanket assemblies in which part of the uranium is converted to fissile material, an additional incentive, in the commission's view, for considering the breeder fuel cycle in future nuclear energy scenarios.

Supply/Demand Conclusions

According to the Conservation Commission, estimates of uranium resources and enrichment requirements may be on the low side for several reasons. Among these, less-developed nations may increase their estimated share of the world's electric power generation capacity leading to increased use of light-water reactors, and nuclear power may penetrate the electrical market by as much as 65%, in contrast to the assumption of 50%. Offsetting these assumptions, however, is the final projection of total world electrical energy demand envisioned by the commission, a projection less than that used in its resources analysis.

The net result of these interacting assumptions, the commission feels, would be a somewhat greater demand for uranium resources. Moreover, while only the requirements for producing electricity were considered, the use of nuclear energy as a heat source, for example, is expected to have applications in such areas as district (central) heating, steel-making, and synthetic fuel production, thus adding to the total demand over the study period. (It appears, however, that prior to the year 2020 the amount of nuclear energy used in such applications is not likely to be as great as the variations attributed to uncertainty in the amount of nuclear energy used in electrical power generation.)

Perhaps one of the more important factors which may tend to depress the projection of uranium requirements is the assumption that the rate of penetration for advanced fuel cycles will be similar to that of the light-water reactors. This assumption is probably optimistic with respect to breeder-reactors and other advanced fuel cycles, and leads to the conclusion that many new techniques will require more or less simultaneous development, thus adding to the complexity of the overall program and to the liklihood that penetration rates will be longer than assumed.

With the single exception of the United States, the commission points out, the bulk of presently-known uranium resources are distributed among countries whose nuclear programs are expected to be relatively small in the pre-2000 period. On the other hand, a rapid increase in nuclear power growth is foreseen after the year 2000 in regions which are expected to yield a large portion of the world's undiscovered uranium resources. This would possibly ameliorate the long-term supply situation.'

The task of providing the world with uranium at the rate required, even under the rather optimistic demand assumptions in the commission's scenario number one, is truly immense. Indeed, the problem is not likely to be one of the existence of resources, but rather of the availability of the required resources on an appropriate time-scale. Of the numerous factors which will make meeting world demand exceedingly difficult, the factor of time is perhaps the most crucial. Under the favorable assumptions of scenario one, the uranium supply industry is being asked to increase its level of production fifteen-fold in less than 45 years. In view of the ever increasing number of constraints, and the lengthening lead times required between initiation of new exploration programs and realization of initial production from successful discoveries, the 45-year period seems all too short.

According to the commission, achieving the projected levels of uranium production is a challenge calling for unprecedented levels of effort and international cooperation.

Nuclear Fusion

The fusion process can be considered a nuclear analogue of chemical combustion. Both types of reactions are accompanied by release of energy, and in each case the "fuel" must be raised to an "ignition" temperature before energy release begins.

However, with fusion both the temperature required for ignition and the amount of energy released per event are greater by many orders of magnitude. In fact, the high temperatures required for fusion reactions—approximately 100 million degrees Celsius for the most easily "combustible" fuel combination—present some of the major problems associated with a practical reactor. First, means must be developed to heat the fuels, which at ignition temperatures are highly ionized gases, commonly called "plasmas." Next, means must be found to maintain the highly energetic fuel nuclei at a sufficiently high density and for a sufficiently long duration for a significant number of reactions to take place before the plasma is dissipated.

The amount of fusion power produced is a function of plasma temperature, the amount of plasma present, and the nature of the fuel mixture. In order to have a useful reactor, the amount of energy released during the thermonuclear burn must exceed by a significant factor the amount of energy required to initiate the reactions. Indeed, for satisfactory operation a large fraction of the fuel must be burned each time it is brought up to reaction temperatures.

There are a number of possible fusion fuel cycles, all of which use lightweight elements—such as the isotopes of hydrogen, helium, lithium, and boron—to produce energy.

The fuel cycle that utilizes deuterium and tritium, both of which are isotopes of hydrogen, has the lowest requirements for plasma temperature and confinement time, the highest probability of occurrence of all the possible fuel cycles, and also releases a significant amount of energy with each reaction. For these reasons, the bulk of fusion research to date has been concentrated on this cycle, and it will almost certainly be used for scientific feasibility demonstration and probably for the first generation of power reactors.

Although problems associated with the fusion process have proven to be considerably more difficult than was originally envisioned some 25 years ago when the quest for harnessing fusion power began in earnest, progress in understanding the behavior of plasmas has been most impressive in the past decade. As a result, while knowledge of the underlying physics is extensive, it is not yet fully developed, and consequently it is impossible to predict with certainty when—or even if—scientific feasibility will be accomplished. Despite this, recent achievements in plasma production, plasma heating, and plasma confinement have led to a widespread belief among the scientific community that this goal will be reached within the next five to ten years by one or more of several approaches under current investigation.

Fusion Energy Resources

Deuterium can be separated from water, and it is estimated there exist in the earth's seas approximately 46,000 gigatonnes of this isotope—a virtually inexhaustible supply. Tritium, although it must be produced artificially, is in much the same class when produced in a breeding process with lithium. This breeding process occurs in a fusion reactor in which lithium—a relatively rare element—absorbs fast neutrons to create the tritium.

Current estimates of the world's recoverable land reserves and resources of lithium (i.e., exclusive of lithium dissolved in the oceans) are about 10 million tonnes. As a result, the resource situation is somewhat like that of thorium since in both instances, at present annual-demand rates, known resources are adequate for hundreds of years. This being the case, exploration activities are minimal and the probability of new discoveries is high. The recovery of lithium from seawater, and particularly solar bitterns, may eventually provide a near-infinite source of supply for this element at a stable price. Estimated lithium in the oceans is approximately 184 gigatonnes.

Energy From Fusion

The fusion energy content of the world's supplies of deuterium and lithium is seemingly limitless. The total energy content of one tonne of deuterium is 0.24 exajoules, and that of one tonne of the lithium isotope is 0.36 exajoules. In addition, there is some indication that achievable energy outputs may be two to three times these amounts depending on reactor design.

If lithium were totally recovered on land and sea, the energy equivalent would be about 66 billion exajoules. At the same time, the energy supply from the complete utilization of the estimated deuterium available in the earth's waters would represent some 11,000 billion exajoules.

Fusion's Versatility

Although major research efforts in the fusion program throughout the world have focused on large, deuterium-tritium based reactors designed to produce electric power, investigation has also begun into alternative fuel systems and uses.

For example, the large neutron and other radiation fields possible with fusion reactors could be used for producing fissile materials for fueling conventional nuclear reactors, for producing chemical fuels, or for the transmutation of radioactive wastes from fission reactors. Research is also being conducted into the possibility of developing reactors of different sizes, configurations, operational characteristics, and fuels. This potential versatility of application could be an important asset of fusion power.

Future Outlook

According to the commission, no measurable amount of energy is expected from fusion by 2020. However, the potential contributions of fusion power are great and could be commanding in in the course of time. A variety of approaches now under active consideration could succeed, or other approaches not yet well explored could prove to be the ultimate favorites. The state of the art is embryonic by comparison with many other alternative energy conversion technologies, and it would be premature to say with any conviction when fusion might begin to enter the commercial market in a serious way.

Nuclear Energy Research & Development

Current research and development on nuclear energy may be classified into seven main categories: uranium exploration and mining; support for existing thermal fission-power reactor systems; advanced fission-power reactor systems (notably the liquid-metal, fast-breeder reactor system); fuel cycles and safeguards; the thorium cycle; fusion; and, spallation and other research.

Uranium Exploration and Mining. Concern about the environment, health and safety is perhaps the major thrust of research and development in uranium mining and ore processing, giving emphasis to the monitoring and control of radon gases underground; the safe disposal of radium-226 in tailings; and to improved leaching techniques to recover both radium and thorium. Since it is expected that uranium deposits will become more difficult to find, several countries are also studying improved exploration techniques, including remote sensing (among them, radiogenic heat sensing) and radiometric instrumentation. Uranium ore genesis and uranium geochemistry are also subjects of study.

Support for Existing Thermal Fission-Power Reactor Systems. Attention here is on improved analyses of steady-state reactor physics and long-term and short-term dynamic performance; studies of fuel and its behavior; development of components; reactor fluid chemistry and system corrosion; accident analyses and related experimentation; the testing of new systems; and control design. New information from operating plants is also being carefully analyzed.

One specific mentioned here is the high-temperature gas-cooled reactor which is receiving attention particularly in the U.S. and the Federal Republic of Germany, and offers good prospects for very efficient generation of electricity employing gas turbines. In addition, this reactor may also have applications for metallurgical purposes and for liquefying coal.

Advanced Fission-Power Reactor Systems. Research related to the fast-breeder reactor—particularly the liquid-metal-cooled version, LMFBR—is attracting the greatest expenditure of development effort in the nuclear energy field. Several countries have major programs, and some, Great Britain, France, the United States, and the U.S.S.R., have operated small pilot units for years. Other countries, such as Japan and India, will soon join them.

In addition, Great Britian, France, and the U.S.S.R., have prototype nuclear electric power plants in the 200 to 300 megawatt class which employ liquid-metal-cooled fast-breeder reactors. The Federal Republic of Germany, with Belgium and The Netherlands, is also building a plant in this class, while the U.S.S.R. is building a 600 megawatt prototype, and France, the Federal Republic of Germany, and Italy are cooperating on a 1,000 megawatt unit.

The attraction of the LMFBR is that it produces more plutonium by conversion from uranium than the amount of uranium and plutonium it consumes. This makes it at least theoretically possible, eventually, to consume virtually all the uranium rather than the 0.5% consumed in thermal reactors of current design. Development of the LMFBR promises an essentially unlimited energy resource, since the markedly better use of available uranium makes the use of poorer ores economic, thereby compounding the advantage. However, many aspects of the LMFBR require major research and development effort.

Fuel Cycles and Safeguards. The design challenge with the LMFBR also involves developing processes for treating the irradiated fuel so as to recover the plutonium and uranium, for separating the waste products of the fission reaction, and for disposing of the waste.

Incorporating the wastes in ceramics, glasses, or mixed matrices, and burial in deep mines in salt or stable hard rock, are the preferred techniques for waste isolation today. Twenty years ago, in fact, experimental quantities of reactor fuel wastes were fixed in glass and buried in wet, sandy soil in Canada. These have been monitored since that time and the evidence is that this would be an adequate disposal technique if there could be assurances that the wastes would not be unwittingly disturbed. The purpose for burying them in deep-mined cavities in stable geological formations, it is pointed out, is to provide that assurance.

According to the commission, a prime interest in fuel cycle development is to produce designs that do not unnecessarily aggravate the risk of exposing fissile material to diversion from a power program to possible destructive uses—and if possible to reduce that risk. An international fuel cycle evaluation study was established for cooperative investigation on this subject in 1977 and is expected to report its findings in 1979.

The Thorium Cycle. In addition to development of the uranium-plutonium fuel cycle for the LMFBR, the United States is conducting research into the thorium-uranium-233 fuel cycle for possible application in lightwater reactors. Although a high rate of conversion of thorium to uranium 233 can be achieved, still better conversion ratios can be achieved in the high-temperature gas-graphite reactor and the heavy-water reactor. In the Canadian version of the heavy-water power reactor it is expected that conversion ratios approaching and even slightly exceeding unity can be achieved. With such performance, energy comparable in amount to that obtainable through the use of the LMFBR could eventually be derived from the world's stocks of uranium. However, since the thorium-uranium-233 cycle cannot—even in the most efficient thermal reactor—generate as much new fissile material as the LMFBR, more uranium would be needed in the early decades of any program in this area.

Fusion. Research into controlled nuclear fusion has been in progress for more than 25 years, and during the past decade interest in development of a fusion reactor for energy production and other purposes has increased significantly. The United States, the U.S.S.R., the Euratom countries, and Japan have major programs, and several other countries have some scientific and technical involvement.

Two general approaches to achieving the conditions required for break-even fusion are being pursued. They are magnetic confinement, in which hot plasma (the cloud of ions) is confined by a powerful static magnetic field, and inertial confinement, in which fuel in tiny spherical form is heated and compressed by convergent powerful beams of laser light or electrons. Two dozen major experimental facilities are in operation, under construction, or planned, to probe toward density-time-temperature combinations conducive to break-even fusion processes.

According to the commission, while it is by no means clear that conventional fusion reactors are a practical objective, it hardly seems credible that with the understanding already achieved, and the record of past accomplishments, controlled nuclear fusion will not one day be mastered. Nuclear fusion represents one of the greatest technological adventures that man has embarked upon and, if successful, it will eventually provide a virtually limitless source of controlled energy.

Spallation and Other Research. The object of the fast-breeder reactor and the thorium fuel cycle is a process that will fission all the heavy element (uranium or thorium) material supplied to it. Of natural materials, however, only atoms of uranium-235, which constitutes 0.7% of the total uranium, have a high probability of fissioning when they receive a single neutron. Other uranium atoms and all thorium atoms usually require one neutron to "cock" them and a second to "fire" them. In the case of uranium, this "cocking" involves the conversion of uranium to plutonium. In the case of thorium, it involves the conversion of thorium to uranium—233.

It is envisioned that the fusion reactor may also be used to "cock" uranium or thorium atoms. Another process known as "spallation" is also being considered. With spallation a proton is introduced into the nucleus of an atom causing a small fraction of the atom to spall, or chip off. Researchers in a number of countries are active in investigation into these processes.

Geothermal Energy

Geothermal energy is a virtually unlimited resource representing energy transferred to water or steam from molten, underground rock which receives its heat from the slow decay of naturally-occurring radioactive elements present in all rocks. Geothermal energy resources are those estimated to be contained in the earth's crust to a level of about five miles, and are more or less concentrated along well-defined belts that cover about 10% of the earth's surface. These belts are associated with continuous earthquake activity and historical volcanism, and also with geological crustal plate boundaries.

Geothermal resources are generally classified into four basic types distinguished by their thermodynamic and hydrologic characteristics: liquid dominated hydrothermal; vapor-dominated hydrothermal; petrothermal; and geopressured reservoirs.

Liquid Dominated Hydrothermal. Geothermal resources are called

Table 14 Estimated geothermal resource base as reported by region

Region	*Energy resource, EJ × 10^3*
North America	1700
Central America	—
South America	0.54
Western Europe	360
Eastern Europe	54
Asia	0.08
Africa	—
Pacific Islands	0.13
Total	2115

liquid-dominated if the pore space in the reservoir is filled with water or a brine solution that is circulated by convection to transfer heat energy from the underlying rock. Typical temperatures in such systems range from near ambient to as high as 360 degrees Centigrade. Exploration has revealed that of the hydrothermal resources near the earth's surface, liquid-dominated systems are far more abundant than vapor-dominated ones.

Vapor-dominated Hydrothermal. Sometimes referred to as "dry steam fields," vapor-dominated systems are relatively rare. However, the most economically favorable geothermal power developments in the world today are associated with vapor-dominated systems. In these systems, saturated steam is the continuous fluid phase in most of the pore space. Most investigators assume that water exists either as pore space droplets or in bulk volume at the bottom of the reservoir. Temperatures typically range from 220 to 250 degrees Centigrade. Production is relatively simple, and a slight superheating of steam often occurs during extraction.

Petrothermal. Since porosity generally decreases and temperature increases with depth, it is likely that vast volumes of hot, dry rock exist at great depths within the earth's crust in both solid and molten forms. Methods are being investigated for introducing cold water into these hot, dry rock systems and then extracting the heated water which results through wells adjacent to the injection wells. Engineering and economic feasibilities, however, have not yet been established for this process.

Geopressured Reservoirs. These reserves are located in deep, sedimentary strata where compaction has taken place over extended periods of time, and where an effective shale seal has formed. Under these conditions, water is squeezed out of the shale matrix into adjacent sand bodies, becoming compressed well beyond the hydrostatic pressure at that depth. These systems are often characterized by higher than normal temperatures (as high as 237 degrees Centigrade in some reservoirs along the Gulf Coast of the United States) with wellhead pressures of 11,000 pounds per square inch, or more. Additionally, geopressured waters or brine solutions typically contain dissolved methane gas, probably to the point of saturation in many cases, and may become an economically important hydrocarbon resource in the future. The amount of natural gas in geopressured zones in the U.S. alone has been estimated at one-third to five times estimated world conventional gas resources. The technology is not yet developed for using geopressured resources, however.

Geothermal Energy Resources

The commission developed its estimates for geothermal energy resources using two approaches. One, a review of published technical literature combined with responses to a survey of member nations of the

World Energy Conference and certain other nations. The second by calculating a resource base from geological and geophysical data.

Commission results indicate that the expected growth of geothermal energy in non-electric applications is less than that for electric applications (due to the fact that transport of low-grade heat over any significant distance is usually uneconomical), and that the predicted supply of energy from geothermal resources might reach 10 exajoules annually by the year 2020.

Electricity from Geothermal Energy

Attempts to generate electricity from geothermal energy date almost to the turn of the twentieth century, with the first experimental geothermal-electric power plant being operated in Italy in 1904 using naturally flowing geothermal steam. In a few other countries, including the United States, Japan and New Zealand, attempts to develop geothermal energy for power production began in the early 1920s but were not commercially successful until the early 1960s. Since that time, geothermal energy developments have been initiated in many countries, reaching a total electrical capacity of about 1325 megawatts in 1976.

Until recently, the incentive for developing geothermal energy was low because fossil fuels seemed abundant and energy planners traditionally were reluctant to depend on an energy resource whose quantity could not be precisely predicted, or whose cost could not be known well in advance. However, since 1973 when worldwide costs for gas and oil rose markedly, reexamination of all energy resources has brought about a renewed interest in geothermal energy for both electrical and non-electrical uses.

Although present overall recovery of potential geothermal energy is rather small, the potential in electrical energy available from geothermal power is truly significant. In fact, in the year 2020 it could potentially exceed today's total annual electrical energy supply of 28 exajoules by 10%.

Future Outlook

The Conservation Commission feels it is likely that geothermal resources could play an increasing role in the world's overall energy supply, and that countries located in major geothermal belts may be able to obtain significant fractions of their energy requirements from this source. Even countries underlain by normal temperature gradients, the commission projects, may find that geological, climatological, and economic circumstances combine to create a favorable economic setting for utilizing subterranean heat.

Hydraulic Energy

Installed hydroelectric generating capacity (electricity generated by water power) stands today at about 372,000 megawatts, with yearly production of 5.7 exajoules (electric)—a total equivalent to 14 exajoules of primary energy. This total represents approximately 23% of the world's electricity, and is equivalent to the use of about 2.5 billion barrels of oil a year in an oil-fired power plant (or some 6.5 million barrels of oil per day).

Man's earliest practical use of energy, other than from fire and the muscle power of animals, was derived from the winds and flowing water. The use of waterwheels of various types extends back to early civilizations when energy was harnessed by mechanical means for use at the site to perform such tasks as grinding grains, pumping water to higher elevations, and sawing timber.

The size and efficiency of waterwheels naturally improved over the years and in the nineteenth century water-powered mills and factories of various types ushered in the industrial age. Peak development of water-wheels was reached at about the middle of nineteenth century. Both site limitations restricting use to riversides suitable for dam and channel (sluice) construction, and development of the more flexible and easily sited steam engines, which improved economy and dependability, served to curtail use.

With the advent of direct-circuit electrical power in the late 1880s, followed by development of alternating-current technology, transmission of electrical energy over relatively long distances became possible. As a result, after 1910 the installation of hydroelectric facilities accelerated in North America, Europe, and Japan, and by 1925 about 40% of the electricity generated in the world was hydroelectric. Installed hydroelectric capacity in 1925 was approximately 26,400 megawatts, with annual production about 0.29 exajoules. This represented about 0.6% of total energy consumed that year.

Hydroelectric developments were rapid in the 1930s with typical installations on the order of 100 megawatts. (The larger Niagara Falls and Wilson Dam projects in the United States were notable exceptions.) After 1930, multi-dam river basin developments came to the fore, and very large dams with hydroelectric plants having capacities of hundreds of megawatts were constructed. Typical of these were dams on the Volga River in the U.S.S.R., and the Hoover and Fort Peck dams in the United States.

Following World War II, the rate of construction of large dams increased in tempo on a worldwide scale with major construction undertaken in Brazil, the People's Republic of China, Egypt, Ghana, India, Pakistan, Turkey, the U.S.S.R., and Zambia-Rhodesia, to name a few.

Today, the greater part of the economically attractive hydro sites in the industrialized nations have been developed, with undeveloped sites in these countries rather remote, and dams and powerhouse construction there quite costly. In less-developed nations, however, there is large potential. Indeed, hydroelectric installations in some areas of Asia, Africa, and South America will provide a significant portion of total energy requirements for those areas.

Development of Hydroelectric Resources

According to the commission, extensive hydroelectric development will occur in the world during the next few decades as a result, primarily, of the fact hydroelectric energy appears to be the most feasible—and in many areas the most economical—of the renewable primary energy sources for electric power generation. This growth will come, it is said, despite growing environmental, legal, and political constraints on site development.

Assuming suitable financing arrangements can be made—a significant problem for less-developed countries—it seems reasonable to expect that those installations reported under construction in a 1976 World Energy Conference survey will be completed by 1985, and further that those installations reported planned in that survey will be completed by the year 2000. This would represent an increase of two-and-a-quarter times present worldwide capacity, and a uniform growth rate of 3.5% over the period 1976 to 2000. This 3.5% rate breaks down to 1.5% in developed nations, and 5.6% for less-developed and centrally-planned nations combined.

Commission findings also show that the limits of installed and installable hydroelectric capacity will be approached in most regions of the world within the first few decades of the twenty-first century arising from pressures to develop renewable energy resources to their maximum. Tending to slow this rate, it is said, will be future demand for water not compatible with maximum power generation (needs such as irrigation of new lands for increased food supplies), and potentially strong environmental pressures in some parts of the world for the preservation of free-flowing streams.

Developed nations represent 24% of the potential world hydraulic resources, the commission reports, centrally-planned nations 28%, and less-developed nations 48%. The greatest potential for development is in less-developed nations where only 7% of projected capacity is in operation. This situation contrasts sharply with the 46% of potential now in operation in developed nations. Estimated probable hydroelectric production in exajoules in the years 1985, 2000, and 2020 is shown on the following page. The figures shown are the probable annual average energy from installed hydroelectric facilities for the year indicated.

According to the commission, production of the 28.3 exajoules (electric) of energy projected for the year 2020 would involve the

Table 15 Estimated probable hydroelectric development

Country groupings	Potential energy in exajoules (1 EJ × 10^{18} J)			
	1976	*1985*	*2000*	*2020*
OECD countries	3.78	4.49	5.37	7.80
Centrally planned countries	0.72	1.20	2.88	8.70
Developing countries	1.17	1.97	4.49	11.80
World total	5.67	7.66	12.74	28.30

The figures shown are the probable annual average energy from installed hydroelectric facilities for the year indicated.

development of 80% of the world's total developable hydroelectric resources, and represent the equivalent of 34.5 million barrels of oil per day (or 1.8 billion tons per year in oil-fired thermal power plants). This is equivalent to 8% of total world energy consumption projected for the year 2020.

Advantages of Hydroelectric Energy

Hydraulic energy has a number of characteristics which make it attractive for future energy supply. Among these are:

- It is a resource which is continuously renewed by the energy of the sun (which creates and sustains the hydraulic cycle);
- It can be an important part of multipurpose utilization of water resources, (i.e., of irrigation, navigation, recreation, and flood control);
- It can be made available in remote areas of less-developed countries, becoming the catalyst in economic development of other resources and improving living conditions; and,
- Its flexibility and dependability make it the most economical and practical of large-scale storage systems.

Factors Affecting Hydroelectric Development

In analyzing the potential for hydroelectric development the commission focused on the economics of development; the multipurpose uses of water; needed and developing technology; environmental concerns; societal concerns; and legal and political concerns.

Economics. The economics of hydraulic power generation improve significantly as costs of fuels for heat-power generation (e.g., oil, natural gas, and coal) increase and supply limitations for these fuels become more apparent. In addition, hydroelectric projects are especially attractive to less-developed nations since such projects represent an immediate technology within the competence of indigenous construction industries, and construction is more labor-intensive than capital-intensive.

Multipurpose Uses of Water. In many areas of the world water supply is the controlling factor of present and future human activity. As a result, a comprehensive master plan for a river basin, perhaps including adjacent river basins, is essential for optimum regional development—especially as water supply becomes a more critical consideration in the region. Indeed, a hydroelectric project can be the economic support for a comprehensive plan for river basin development.

Technology. Advances in technology will have a real impact on development of hydroelectricity. As an example, a better understanding of hydrology is needed to help anticipate and prepare for fluctuations in rainfall and water supply. In addition, expansion of remote-control capabilities for hydroelectric plants will make for more efficient use of energy, and the increasing ability to transmit high voltages—perhaps over a million volts—over long distances means that an increasing number of large and remote hydroelectric projects can be constructed to serve distant population centers.

Environmental. Hydroelectric development on rivers and streams can have both positive and negative impacts. Specifically, consideration should be given to the effect of such development on biological systems dependent on the water resource, on flood prevention, silt and debris retention, and on recreational and other human benefits.

Societal. Hydroelectric installations can initiate changes which will bring primitive, poverty-stricken cultures a significantly improved living standard. At the same time, however, large developments may require the relocation of villages and large numbers of people.

Legal and Political. Legal rights to water vary widely with some areas having essential property rights, others common-law riparian rights, and still others little or no legal framework. These concerns become especially important in international ventures, a number of which have been undertaken worldwide, requiring significant international cooperation.

Future Outlook

The Conservation Commission believes that although hydraulic energy's share of the electrical market will fall to about 16% by 2020 from 23% today, it will in fact be of greater relative importance as an energy source. Its advantages include the following:

- It is continually renewable and virtually non-polluting;
- It is an integral part of optimum overall water resources utilization, and continues to be an important capacity component of large electric power systems because of its reliability and flexibility;
- Its development often serves as the initiating catalyst to economic improvement in less-developed nations; and

• Its economic position relative to other electric power generating facilities will continue to improve because it has a non-inflationary energy component (water) in its overall generating expense, and it is characterized by long-lived plant with low maintenance expense.

As regards future development of hydroelectric resources, the commission concludes that in less-developed nations capital problems encountered in launching construction should be alleviated by the increasing economic need for, and value of, hydroelectric power. In developed nations, it indicates that development should continue with speical attention being given to expanding generating facilities at existing reservoirs and installing them at dams where they do not exist today.

Solar Energy

Solar energy is often thought of only in terms of the energy potential of direct sunlight. However, indirect forms of solar energy—such as wind energy, wave energy, tidal energy, ocean thermal-energy conversion, and biomass energy (firewood, waste, and animal dung)—are also important and potentially very significant.

According to the Conservation Commission, the amount of solar radiation arriving at the outer boundaries of the earth's atmosphere is equivalent to five million exajoules per year, about 65% of which appears in the form of terrestrial heat, the evaporation-precipitation cycle, wind, wave, convection, and ocean currents, and photosynthesis. Gravitational energy—including gravitational resources, tides, and tidal currents, all of which have their origins in the potential and kinetic energy of the sun-moon-earth system—is estimated to be on the order of 100 exajoules a year.

Overall, solar energy reaching the earth's surface is 13,000 times greater than the total present world production of energy from fossil fuels and uranium.

While this is a staggering total, in the commission's assessment some four decades will be required before solar energy systems can be commercialized. Indeed, such systems are said at present to be in so early a stage of development that forecasts of their potential contribution toward world energy supply within the next 40 to 50 years would be little more than conjecture.

One example of solar technology which has reached a fairly mature stage of technological development but has yet to demonstrate economic advantages on a broad scale is the use of flat-plate collectors to convert energy from the sun to low-grade heat for supplying hot water for residential use, or heat for residential and commerical structures. Another example, the conversion of solar energy to electricity, is known to be both scientifically and technically feasible, but this is not yet far enough advanced to judge whether it can be economically attractive in comparison with alternate means for electrical generation.

Despite this uncertainty, a projection for sources and applications of solar energy has been attempted for the year 2020. It shows the following (note—energy supplied at the rate of one terawatt over a period of one year is equal to 31.54 exajoules):

Table 16 Solar energy sources

Source of application	*Global abundance, TW**	*Theoretically recoverable, TW†*	*Range of contribution by 2020, TW‡*
Solar	178 000	50—100	
Heating/cooling			0.5—2.0
Electricity			0.1—0.3
Fuel			?
Wind	350	?	
Electricity			<0.01
Mechanical work			<0.20
Waves	3	0.1—1.0	
Electricity			<0.01
Fuels			
Tides	3	0.06	
Electricity			<0.01

* Crude estimate. † Assuming appropriate technology can be developed economically. ‡ Assuming current projections for near-future technological successes are accurate.
Energy supplied at the rate of one terawatt (TW) over a period of one year is equal to 31.54 exajoules (1 EJ × 10^{18} J).

According to the commission, in order to appreciate the relative magnitude of potential contributions from solar energy in the time-frame 1985 to 2020, it should be recalled that the current annual average energy consumption from all sources, worldwide, is in the range of eight to 10 terawatts and is likely to grow to 30 to 40 terawatts by 2020.

As an energy source, solar energy has numerous advantages. It is universally distributed, inexhaustible, and available in the same form over an indefinite time scale. In addition, it is almost wholly environmentally benign.

At the same time, however, it has its drawbacks. Unlike fossil fuels, uranium and geothermal energy, sunlight is not a static deposit of energy, but rather energy in flux. In addition, its characteristics create great difficulties for reliable and economical application. The diurnal, seasonal, and climatic variations in available sunlight, as an example, require the use of sizeable energy storage systems, or hybrid applications with other energy resources in order to provide reliable, round-the-clock energy supply. In addition, solar energy is a diffuse source, varying in intensity from place to place over the earth and therefore requiring elaborate collection and concentration systems when used on a large scale. These conditions place significant restrictions on the amount of solar energy that may be converted to useful energy.

Solar Energy Resources and Utilization Systems

In its report the Conservation Commission discusses five potential systems for solar energy utilization: solar-generated thermal energy; photovoltaic (solar radiant) energy conversion; solar hydrogen production; indirect solar energy (wind, wave, tidal, and ocean-thermal energy); and biomass conversion.

Solar-Generated Thermal Energy. Solar radiation clearly lends itself to the production of heat, with attainable temperatures depending on the design of collection devices. While the commission focuses on heat at less than 100 degrees Centigrade, which is readily achievable without concentration, the theoretical maximum temperature attainable by concentration is approximately 5600 degrees Centrigrade. A solar furnace at Odeillo, France, has in fact achieved temperatures approaching 3800 degrees Centrigrade.

The generation of such low-temperature heat (less than 100° C.) from sunlight is a fairly mature technology with a long history of application in many parts of the world. The basic system consists of a flat-plate collector containing a liquid or air heat-transfer medium, with or without storage, used for water and space heating. (Air conditioning generally requires somewhat higher operating temperatures.)

The commission estimates that simple applications of solar energy to water and space (air) conditioning are not likely to displace more than 2 to 5% of the world's primary energy requirements, and are unlikely to reach even this range by the year 2020.

A promising technology incorporating solar-generated thermal energy is the "solar thermal-electric conversion" system for producing electricity, a technology which could also be used to produce higher-grade heat needed for a variety of industrial processes. Basic elements of such systems include an optical concentrator to gather solar radiation, an absorber and storage system for the thermal energy collected, and a heat-transfer system linked to a more or less conventional electricity generating plant. In principle, power plants of this type can come in a variety of sizes ranging from a few hundred megawatts down to very small units.

In the United States, an active solar thermal-electric system development program is being pursued with plans for the construction of units varying in size from a 10 megawatt (electric) pilot plant to a 100 megawatt (electric) unit to be constructed by the mid-1980s. In Europe, two firms are offering such units in sizes up to one megawatt (electric), and additional construction plans have recently been announced by France and Spain. In Japan, a pilot plant with two units of one megawatt (electric) each, each being a different type, are to be constructed.

One important fact in considering the development of solar thermal-electric conversion systems, however, is that to date the use of such systems for base-load generation has received relatively little attention. In addition, the reliability of energy supplied by this means will be significantly influenced by the siting strategy chosen and the nature of the total utility system with which it would be interconnected.

Photovoltaic Energy Conversion. Solar radiant energy may be converted to a direct-current electricity by means of suitably constructed thin layers of semiconducting material. For the past two decades, in fact, the silicon solar cell developed in 1955 by Bell Laboratories in the United States has been the mainstay of spacecraft power systems.

The potential advantages of photovoltaic energy conversion include: there are no moving parts, lifetimes can be very long, maintenance involves little skill, both direct-beam and diffuse solar radiation can be utilized effectively, and the system is inherently modular readily lending itself to the design of virtually any system large or small.

Present costs of solar cells range from $10,000 to $15,000 per peak kilowatt of electricity—some 50 to 100 times as expensive as conventional methods—and have proven efficiencies of 12 to 15% (as compared with efficiency limits of about 23%).

According to the commission, while alternatives to silicon may reduce overall costs of the solar cell, wide-scale use of photovoltaic devices for power generation will require a number of other conditions. Among them: major cost reductions in cell fabrication, proven reliability, compatibility with the remainder of the electric grid system, economic and reliable energy storage, and high-power soldi-state regulators and inverters for converting direct current to stable alternating current.

If either or both the solar thermal-electric conversion system and photovoltaic technology can begin to enter the electricity generation market economically by the 1990s, a significant fraction of the electricity produced globally by the end of the year 2020—perhaps 10 to 15%—could be furnished by these means. At this stage, however, such estimates are highly conjectural.

Solar Hydrogen Production. Hydrogen gas, either individually or in a related compound such as methanol, is a secondary form of energy that can be readily produced by solar energy. While little serious effort has been devoted to such production to date, present estimates for the cost of producing hydrogen by solar-chemical means range from $40 to $80 per barrel of oil-equivalent.

Problems here center on the large solar installations that would be required. These problems include the facts such installations would have to be in the sunny, arid regions of the world, regions which are inhospitable for any other human activity; they would require pipeline and tanker transport and distribution networks to the world's energy markets; and large-scale storage challenges would be presented. While the commission feels it would be premature to estimate the potential contribution of hydrogen to world energy demand in the year 2020, it concludes the option deserves continued attention and expanded research and development.

Indirect Solar Energy. The commission's discussion of indirect solar energy focuses on wind, wave, and tidal energy, and ocean thermal-energy conversion.

Wind Energy. Until it was displaced by the steam engine, wind energy had historically been used on a large scale for commercial and agricultural purposes in various European nations. Today, however, development efforts are being focused on applying it to the generation of electricity using large generators linked with existing electrical networks.

There are a number of potential limitations to this electrical generating capability, however, each of which must be evaluated before a meaningful assessment of wind power's role can be made. Among these are: economic considerations likely to limit the use of wind generators to locations with mean windspeeds greater than 12 miles an hour; the fact of diurnal and seasonal wind variations; problems of diurnal and seasonal wind variations and site accessibility; and problems of interconnection to a utility network. Tied to these is the fact more information is needed about windspeed and its duration at 70 to 100 feet above ground level at various locations throughout the world.

At present, the simplest mode for the use of wind power on a substantial scale is by connecting to a grid with existing hydro-capacity where energy may be stored by displacing water. In this application it is projected that despite the fact wind generators today cost two to four times more than conventional generators of equivalent rating, wind power may become a significant energy source on a regional basis where windspeeds and hydrocapacity are particularly favorable.

Overall, according to the commission, the amount of energy potentially available from the wind is probably only a small percentage of that from direct solar energy.

Wave Energy. In spite of extensive human experience with ocean waves over the course of thousands of years, relatively few attempts have been made to exploit this power source for useful energy production. According to estimates, the amount of energy stored in waves (energy derived from wind energy, which in turn derives from the sun) is about 85 exajoules.

The challenge in harnessing wave energy is to find a cost-effective way of converting a fraction of the energy in the wave motion to a mechanical, hydraulic, or pneumatic form that may be used to generate substantial amounts of electric power. Not only must this be accomplished under the inhospitable conditions of the open seas, but also—inasmuch as waves come in a wide range of wavelengths and amplitudes—any effective harnessing device will have to be either broadband (non-resonant) or have its resonance frequency continuously adjusted. Many mechanisms have been proposed for this and a number are being investigated.

According to the commission, worldwide application of wave energy conversion systems will be limited by geographic factors and the economics of energy transport and distribution over long distances. While it is difficult to estimate the potential energy from this source, it too—as with land-based wind-powered systems—is probably but a small percentage of that potentially available from direct solar energy.

Tidal Energy. Two tidal-power generating stations are being commercially operated today, one in France, and one— a small experimental unit—in the U.S.S.R. Although discussions of the prospects for the use of tidal power have extended over many years, the total resource base is small. It has been estimated that no more than 25 sites worldwide meet the necessary conditions of tidal amplitude and coastal topography. Ignoring engineering limitations, this represents an average hydraulic power of about 2 exajoules, of which no more than 10 to 25% could be converted to electricity. In practice, therefore, it is unlikely that more than a small percentage of the total would ever be harvested usefully.

Ocean Thermal-Energy Conversion. The concept that the temperature difference between the surface and the deep waters of the ocean could be used as a source of energy reaches back nearly a century. Recent preliminary technical and economic analyses, however, lead to the general conclusion that much work is required to overcome the substantial technical problems and unfavorable economics involved—problems so large they are expected to attract little research and developement effort.

According to the commission, the engineering problems are so complex with so little margin to maneuver that the potential for this energy form must be considered somewhat speculative at this time. One of the major problems would be to transport the power generated over or in the sea to land-based load centers. As a result, ocean thermal-energy conversion technology does not promise technical and economic viability comparable to the solar-thermal conversion and photovoltaic systems.

Biomass Conversion. Biomass—firewood, waste (e.g., urban and agricultural), and animal dung—is today a significant energy resource, representing 10% of the world's primary energy consumption. Rural populations of less-developed nations, it is estimated, may presently be using up to 31 exajoules (thermal) of energy in the form of biomass. In the future, these less-developed nations are expected to use biomass in an equally great role in the generation of process heat and synthetic fuels.

Of the biomass forms, wood is the most significant, with the photosynthetic production rate of the world's forests producing perhaps 1250 to 1550 exajoules (thermal) of energy—or roughly ten times total present world production of oil and natural gas. Until the early 1800s, in fact, the conversion of sunlight to energy in the form of wood for combustion was the major energy source in much of what is now the industrialized world. what is now the industrialized world.

In its work the Conservation Commission analyzed the role of biomass separately for less-developed and industrialized nations.

Less-Developed Nations. The use of biomass as a fuel source in the less-developed nations is a matter for urgent attention, the commission reports. The problem is two-fold. First, increasing demand for firewood has caused real pressure on forests, making deforestation on a globally significant scale plausible. In much of the less-developed world, it is pointed out, the purchase of firewood is now regarded as one of the major costs of living. Second, combustion of dung and agricultural waste robs the soil of important nutrients and humus. The combined effect of these processes, the commission says, could be to exhaust both firewood and soil fertility leaving no means to replenish either.

An additional problem in less-developed countries is low energy efficiency. Indeed, much of the firewood, dung and agricultural wastes used for energy in these regions is converted to work at far lower efficiency than primary resources in the industrialized world. (Normally, it is said, these fuels are converted to useful energy for cooking, heating, irrigation, and other agricultural tasks with efficiencies of 3 to 5%. This compares with around 20 to 35% for modern fuels and energy-using equipment.) Furthermore, in agriculture (perhaps the most critical resource-based sector in the less-developed world), energy is now the central issue. The need for energy for pumped irrigation and the production of fertilizers alone requires a substantial increase in the efficient use of a far greater amount of energy than is currently available to the less-developed world.

According to the commission, a thoughtfully designed program for biomass conversion offers the promise of turning the inefficient and destructive pattern of utilization of wood, dung, and agricultural wastes into a regenerative system. Indicative of this is the fact gas of an intermediate quality and residue that provides an excellent fertilizer may now be produced from animal dung. This gas, it is said, answers the need for domestic heating and cooking and fertilizer requirements, offering the possibility of electrical generation for lighting as well.

Industrialized Nations. Urban wastes, woodchips, and agricultural wastes represent important sources of biomass in industrialized nations. Beyond these, however, a class of plants called Euphorbia has been proposed. Closely related to the rubber tree and producing a hydrocarbon somewhat similar to petroleum, this class of plants could conceivably produce hydrocarbon fuels at $1 per gigajoule, about half the world market price. Such synthetic fuels could be used in power plants or transformed by chemical processes to methanol or methane to be used in the transportation or petro-chemical industries.

According to the commission, biomass conversion in the industrialized nations could in principle be a substantial source of energy where adequate land is available. Some investigators have even estimated that the worldwide production of fuels from biomass could be as much as 315 exajoules (thermal), a level which would require a growing surface on the order of two to three percent of the earth's total land area. Accomplishing this level is doubtful, however, given the already strong pressures on land for the production of food, fiber, and forest products—pressures expected to intensify in the future—and the practical limits of water and fertilizers.

Hydrogen

Hydrogen is receiving attention as an energy form for several reasons. In the first place, it is an essential component of hydrocarbons, that great family of chemicals that serves mankind as food, fuel, and fiber. Second, as a gas, it can be transported through existing pipeline networks extending the useful life of this piece of infrastructure when natural-gas supplies begin to dwindle. Third, it would be a versatile and valuable store for energy from a cyclical source such as solar energy, or from a capital-intensive source operating on a partial-duty cycle, such as nuclear-electric power generation. And last, it is not a fossil fuel but an abundant, recyclable resource.

Two general methods of hydrogen generation are being explored: they are thermochemical processes and electrolysis.

In the former, water is split by processes involving heat and hydrogenous compounds, such as the catalytic pyrolysis of hydrogen iodide and sulphuric acid. The cations—in this case the iodine and sulphur trioxide—and the catalyst, are recycled. Dozens of possible processes have been analyzed. The most promising offer efficiences in excess of 40% and a few are being developed to pilot-plant stage.

Electrolysis is the classic method of producing hydrogen and the search here is for combinations of electrolytes, electrodes, separators, and operating conditions that might offer the prospect of large-scale economic hydrogen production. Electrolytes include acid solid polymer systems as well as the conventional alkaline system. Oxygen-electrode materials are being investigated as alternatives to nickel. Several materials such as potassium titanate fibers and polytetrafluoroethylene are mentioned for investigation as possible separator materials. Greater efficiency can be achieved with operation at higher pressure and temperature.

According to the commission, it seems inevitable that hydrogen will one day be a major energy and industrial commodity. However, this depends on the cost of natural and synthetic hydrocarbons, specific demands for hydrogen (e.g., in metallurgical processes), the development of energy systems (e.g., solar and nuclear) that could benefit from energy storage, and progress in the development of hydrogen generating systems.

The Conservation Commission predicts that as the margin narrows between the costs of alternative energy and material supplies and the prospective cost of hydrogen, the rate of relevant development will increase. Hydrogen as an energy resource will probably become a major field of energy research and development in a couple of decades, the commission projects.